Principles of Geographical Information Systems

Principles of Geographical Information Systems

Melissa Walker

STATES
ACADEMIC PRESS
www.statesacademicpress.com

States Academic Press,
109 South 5th Street,
Brooklyn, NY 11249, USA

ISBN: 978-1-63989-437-6

Cataloging-in-Publication Data

Principles of geographical information systems / Melissa Walker.
p. cm.
Includes bibliographical references and index.
ISBN 978-1-63989-437-6
1. Geographic information systems. 2. Information storage and retrieval systems--Geography. I. Walker, Melissa.

G70.212 .P75 2022
910.285--dc23

For information on all States Academic Press publications visit our website at www.statesacademicpress.com

STATES
ACADEMIC PRESS

Contents

Preface

Geographical information systems are technological frameworks which are used to gather, analyze and manage data related to positions on the surface of the Earth. It is an interdisciplinary field which draws on the principles of engineering, management, logistics, business and telecommunication. Geographical information system apps are tools which are used for storing and editing spatial and non-spatial data. These can also be used for analyzing spatial information, and sharing the results in the form of maps. Some of the important aspects of this discipline are data analysis, hydrological modeling, topological modeling, geometric networks, cartographic modeling, geostatistics, data mining and geocoding. Geographical information systems find extensive applications across various fields such as wildlife management, urban planning, sustainable development, archaeology, oceanography, economics, etc. Such selected concepts that redefine geographical information systems have been presented in this book. It provides comprehensive insights into this field. This book is meant for students who are looking for an elaborate reference text on geographical information systems.

A short introduction to every chapter is written below to provide an overview of the content of the book:

Chapter 1 - A computer system that captures, accumulates, analyzes and displays data about places on Earth's surface is known as a geographic information system. This chapter will introduce briefly all the significant aspects of geographic information system such as its components, concepts, data structure, database creation and modeling; **Chapter 2** - The process of scanning and recording the physical properties of an object or phenomena without making physical contact with the object is known as remote sensing. This chapter discusses some of the important topics related to remote sensing like its platforms and sensors and remote sensing in solar reflection region. These topics are crucial for a complete understanding of the subject; **Chapter 3** - The art and practice of representing a geographical area on a map or chart is known as cartography. In order to completely understand cartography, it is necessary to understand the processes related to it. The following chapter elucidates the varied processes and mechanisms associated with this area of study like map projection, map designing and map algebra; **Chapter 4** - A standard which is used to encode the geographical information into a computer file is known as a GIS file format. This chapter explains the two types of GIS file formats, GIS raster file formats and GIS vector file formats. All the diverse facts of GIS file formats have been carefully analyzed in this chapter; **Chapter 5** - The database that is developed to accumulate and analyze data representing objects is known as spatial database. This chapter elucidates spatial database management system, spatial data input and editing, spatial analysis and network analysis. The topics elaborated in this chapter will help in gaining a better perspective about database analysis and management; **Chapter 6** - Geographic information system is an interdisciplinary subject which is essential for the understanding of related fields. This chapter discusses the significance of GIS in other sectors like public health, aquatic science, hydrology and environmental governance. The aim of this chapter is to explore the allied fields of geographic information system.

I extend my sincere thanks to the publisher for considering me worthy of this task. Finally, I thank my family for being a source of support and help.

Melissa Walker

1

Geographic Information System: An Introduction

A computer system that captures, accumulates, analyzes and displays data about places on Earth's surface is known as a geographic information system. This chapter will introduce briefly all the significant aspects of geographic information system such as its components, concepts, data structure, database creation and modeling.

Every decision that we take today is largely driven by location and has spatial connotation directly or indirectly. Here about Geographic Information System (GIS) a modern day spatial decision making system. GIS has come a long way from digital mapping to decision making, invoking simulations, creating networks, evolving models based on big geo- data analytics and all gearing for real time decision making with prediction capability. That is where lies the power of geospatial domain and of which Geographic Information System is an important part. The users of this technology have multiplied manifold from few specialists in Universities, space research centres, government departments, and industry to a lay man. The usefulness of GIS is driven today by many pull and push factors and the rapid changes taking place around natural ecosystem. To meet the ever increasing demand created by huge population on infrastructure development, rural and urban planning, safe neighbourhood, multiple transportation facilities, disaster readiness of the community, conservation of nature require a holistic approach with careful understanding of space, created geography and /or varied dimensions of geographical aspect of the earth. Availability of multiple datasets collected from ground platform, aerial platform and space platform have given a new meaning and understanding of the existing locations. Use of GIS capturing various layers of these data sets have definitely added value to the usefulness of the decisions taken by the respective experts of the varied fields.

GIS is a discipline generating massive interest worldwide. It is an interdisciplinary area used by heterogeneous groups what you find is that since mid-1950's GIS as a subject has evolved changes taking place in "theoretical framework, technological improvements and institutional involvement in organizational improvement and inclusive nature" in the GIS field. It has a capability of using large data sets, performing multiple geo-analytics and at the same time geo-visualizing it as per the requirement. But all this begins with geo-referencing–tagging different locations.

Geographic information System (GIS) as a system answers through visualization and data analytics. It helps us to understand spatial relationships, different patterns and changing trends. GIS therefore answers questions related to Location – 'What is at?'; Condition Where is it?; What type is it?; Trends: What has changed since? Patterns: What spatial Patterns exist? And questions related to Modelling-What if? The detailed examples of all the raised questions would be undertaken in subsequent lesson.

The strength of GIS is its ability to integrate multiple data options considering both spatial data with non-spatial data (descriptive information about the spatial data). It has advantages over other similar system which are listed down below:

- GIS visualises where, when, what and why possibilities which is just not feasible with any other software system.
- GIS enables to create a comprehensive framework to a concern in hand. It supports the policy and decision makers to arrive at the right ground reality.
- GIS enable cost saving from greater efficiency; better decision making through improved communication and record keeping.
- GIS predict events/outcomes derived based on multi-criteria analyses carried out using the appropriate thematic layers in GIS.
- GIS assists in planning strategies, especially useful in case where disastrous events demands quick decision making.
- GIS is also becoming essential in designing and prescribing actions in different regions after understanding what is happening and what will happen in geographical space.
- Moreover, through different combination of data sets and geo-analytics GIS generates new ideas never before thought of using earlier studies.
- Above all, GIS geo-visualizes the scenarios given the current situation, how the simulation would look like and its consequences. Such scenarios greatly assist disaster managers to plan for future and safe human lives.

With GIS technology, people can compare the locations of different things in order to discover how they relate to each other. For example, using GIS, a single map could include sites that produce pollution, such as factories, and sites that are sensitive to pollution, such as wetlands and rivers. Such a map would help people determine where water supplies are most at risk.

Data Capture

Data Formats

GIS applications include both hardware and software systems. These applications may include cartographic data, photographic data, digital data, or data in spreadsheets.

Cartographic data are already in map form, and may include such information as the location of rivers, roads, hills, and valleys. Cartographic data may also include survey data and mapping information that can be directly entered into a GIS.

Photographic interpretation is a major part of GIS. Photo interpretation involves analysing aerial photographs and assessing the features that appear. Digital data can also be entered into GIS. An example of this kind of information is computer data collected by satellites that show land use—the location of farms, towns, and forests. Remote sensing provides another tool that can be integrated into a GIS. Remote sensing includes imagery and other data collected from satellites, balloons, and drones. Finally, GIS can also include data in table or spreadsheets form, such as population demographics. Demographics can range from age, income, and ethnicity to recent purchases and internet browsing preferences.

GIS technology allows all these different types of information, no matter their source or original format, to be overlaid on top of one another on a single map. GIS uses location as the key index variable to relate these seemingly unrelated data. Putting information into GIS is called data capture. Data that are already in digital form, such as most tables and images taken by satellites, can simply be uploaded into GIS. Maps, however, must first be scanned, or converted to digital format.

The two major types of GIS file formats are raster and vector. Raster formats are grids of cells or pixels. Raster formats are useful for storing GIS data that vary, such as elevation or satellite imagery. Vector formats are polygons that use points (called nodes) and lines. Vector formats are useful for storing GIS data with firm borders, such as school districts or streets.

Spatial Relationships

GIS technology can be used to display spatial relationships and linear networks. Spatial relationships may display topography, such as agricultural fields and streams. They may also display land-use patterns, such as the location of parks and housing complexes. Linear networks, sometimes called geometric networks, are often represented by roads, rivers, and public utility grids in a GIS. A line on a map may indicate a road or highway. With GIS layers, however, that road may indicate the boundary of a school district, Public Park, or other demographic or land-use area. Using diverse data capture, the linear network of a river may be mapped on a GIS to indicate the stream flow of different tributaries.

GIS must make the information from all the various maps and sources align, so they fit together on the same scale. A scale is the relationship between the distance on a map and the actual distance on Earth. Often, GIS must manipulate data because different maps have different projections. A projection is the method of transferring information from Earth's curved surface to a flat piece of paper or computer screen. Different types of projections accomplish this task in different ways, but all result in some distortion. To transfer a curved, three-dimensional shape onto a flat surface inevitably requires stretching some parts and squeezing others.

A world map can show either the correct sizes of countries or their correct shapes, but it can't do both. GIS takes data from maps that were made using different projections and combines them so all the information can be displayed using one common projection.

GIS Maps

Once all the desired data have been entered into a GIS system, they can be combined to produce a wide variety of individual maps, depending on which data layers are included. One of the most common uses of GIS technology involves comparing natural features with human activity. For instance, GIS maps can display what man-made features are near certain natural features, such as which homes and businesses are in areas prone to flooding.

GIS technology also allows users to "dig deep" in a specific area with many kinds of information. Maps of a single city or neighborhood can relate such information as average income, book sales, or voting patterns. Any GIS data layer can be added or subtracted to the same map. GIS maps can be used to show information about numbers and density. For example, GIS can show how many doctors there are in a neighborhood compared with the area's population.

With GIS technology, researchers can also look at change over time. They can use satellite data to study topics such as the advance and retreat of ice cover in Polar Regions, and how that coverage has changed through time. A police precinct might study changes in crime data to help determine where to assign officers. One important use of time-based GIS technology involves creating time-lapse photography that shows processes occurring over large areas and long periods of time. For example, data showing the movement of fluid in ocean or air currents help scientists better understand how moisture and heat energy move around the globe.

GIS technology sometimes allows users to access further information about specific areas on a map. A person can point to a spot on a digital map to find other information stored in the GIS about that location. For example, a user might click on a school to find how many students are enrolled, how many students there are per teacher, or what sports facilities the school has.

GIS systems are often used to produce three-dimensional images. This is useful, for example, to geologists studying earthquake faults. GIS technology makes updating maps much easier than updating maps created manually. Updated data can simply be added to the existing GIS program. A new map can then be printed or displayed on screen. This skips the traditional process of drawing a map, which can be time-consuming and expensive.

GIS Jobs

People working in many different fields use GIS technology. GIS technology can be used for scientific investigations, resource management, and development planning. Many retail businesses use GIS to help them determine where to locate a new store. Marketing companies use GIS to decide to whom to market stores and restaurants, and where that marketing should be.

Scientists use GIS to compare population statistics to resources such as drinking water. Biologists use GIS to track animal-migration patterns. City, state, or federal officials use GIS to help plan their response in the case of a natural disaster such as an earthquake or hurricane. GIS maps can show these officials what neighborhoods are most in danger, where to locate emergency shelters, and what routes people should take to reach safety.

Engineers use GIS technology to support the design, implementation, and management of communication networks for the phones we use, as well as the infrastructure necessary for internet connectivity. Other engineers may use GIS to develop road networks and transportation infrastructure. There is no limit to the kind of information that can be analyzed using GIS technology.

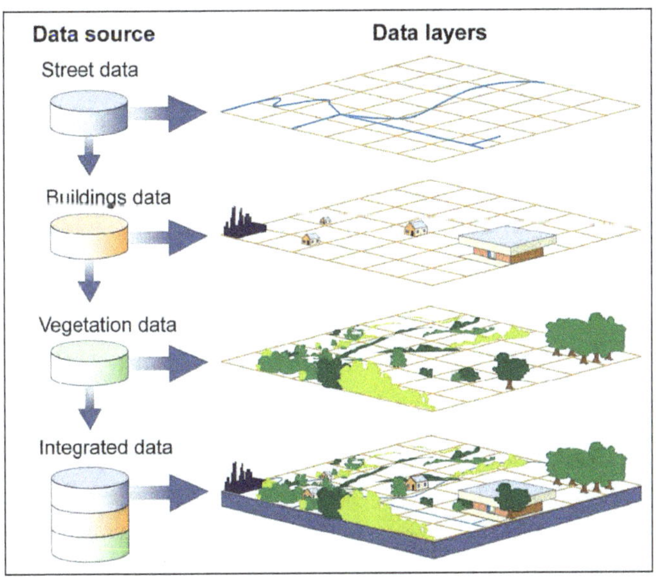

GIS allows multiple layers of information.

Geographic Information

Geographic stands for spatial data. This data must be referenced in space, i.e., geo-referenced and displayed in the form of a 'map'. Data and Information have the three different dimensions namely, theme, time and location. If the source of information or data acquired has a spatial reference, the information can be called as geographic information. Example: Rainfall, Temperature, Elevation of a point above MSL etc. GIS has a definite objective (decision) and information with geographic reference.

The three types of geographic information are:

- Spatial Information
- Attribute (descriptive) information
- Time

Spatial Information

Spatial information pertains to location and shape of geographic features and their spatial relationships to other features. Locational features such as points, lines and areas represent spatial information.

- Point feature: A point may be used to represent a discrete location defining a map object whose boundary or shape is too small to be shown as a line or area feature.
- Line feature: A line is a set of co-ordinates that, when connected may represent the linear shape of a map object too narrow to be displayed as an area (like a river, road etc).
- Area feature: An area is a closed figure whose boundary encloses a homogeneous area (such as a country, forest, water body etc).

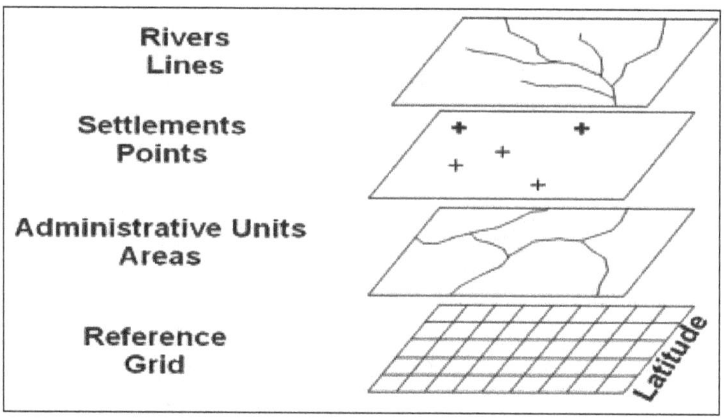

Attribute Information

Descriptive attributes with the spatial features form another important dimension of geographic information. Attribute information contains the descriptions, classifications and measurements of the geographic features. Each geographic feature may have many attributes depending on the significance of it. The attribute data has descriptive, quantitative and qualitative information.

Descriptive information will have elements which cannot be compared with each other in quality and quantity (examples are property ID number, owner's name, vegetation type, zoning type, material type etc). The quantitative information provides for attribute that can be distinguished by measurement and numbers (examples are length, area, value of the property, average age, bearing capacity of soil, voltage etc). The qualitative information forms the attribute data of the features such as water, soil quality and slope.

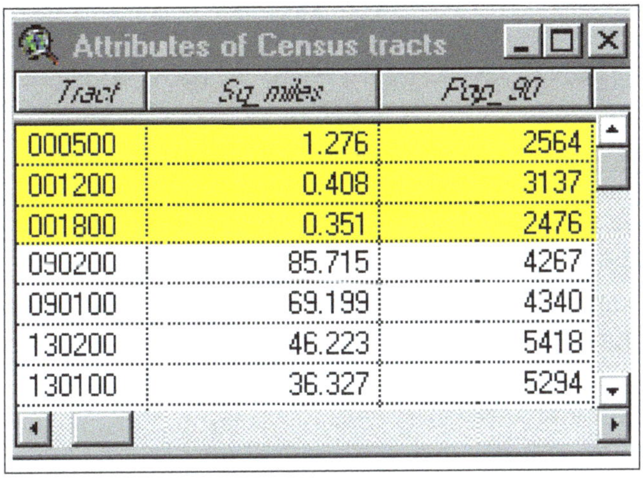
Attributes of Census tracts

Tract	Sq_miles	Pop_90
000500	1.276	2564
001200	0.408	3137
001800	0.351	2476
090200	85.715	4267
090100	69.199	4340
130200	46.223	5418
130100	36.327	5294

Time

With time the geographic information may change. Hence, 'Time' should also be considered as a dimension with the above two dimensions. This is essential to monitor the change in locations of objects and change in attributes.

Classification of Geographic Information

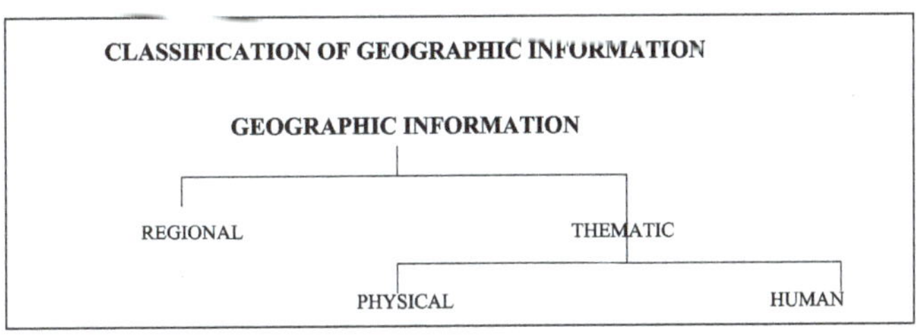

- Regional information modules more than one kind of geographic information in a place or region.
- Thematic geographic information may be one earth feature or human activity as it occurs throughout a wide area. Thematic information includes physical and geographic information.
- Physical geographic information includes the location of such earth features as land, water and climate and their relations to one another and to human activities. Physical features include landforms, climatic patterns, vegetation distribution and soil types.

- Human geographic information concerns patterns of human activity and their relationships with the environment such as political, cultural, population, social, economic, historical and constructions.

Components of Geographic Information System

GIS can be viewed as a software package with various tools to enter, manipulate, analyse and output of geographical data. At the other extreme, GIS components include the computer hardware, software, spatial data, data management and analysis procedures and the peoples to operate it. If the computer is located on a network, it can also be considered as the component of GIS since it enables data sharing among users. Hence, GIS is the combination of all these six components organized to automate, manage, and deliver information through geographic presentation.

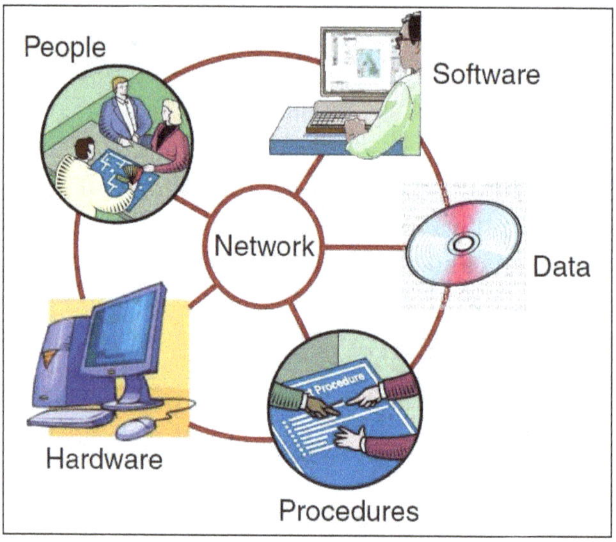

Component of GIS adapted from.

Network

Today, the most fundamental component of GIS is probably the network. Without rapid development of IT, the network, there is no rapid communication or sharing of digital information could occur, except between a small group of people crowded around a computer monitor. Users connected to the Internet could zoom in to parts of the map, or pan to other parts in their desktop WWW browser without ever needing to install specialized software or download large amounts of data. The Internet is increasingly integrated into the GIS. The rapidly growing application of Web GIS has the following capabilities:

- Displaying static maps which users can pan or zoom whilst online.

- Creating user-defined maps online which are in turn used to generate reports and new maps from data on a server.
- Integrating users 'local data with data from the Internet.
- Providing data that are kept secure at the server site.
- Providing maps through high-speed intranets within organizations.
- Providing maps and data across the Internet to a global audience.

Hardware Components

The hardware components of a GIS consist of a computer, memory (CPU or workstations), data storage devices, tape drives or others, scanners, digitizer, plotter, printers, global positioning system (GPS) units, and other physical components. The user controls the computer and the peripherals via a visual display unit (VDU) or terminal.

- The disk drive unit provides space for storing of map and document data in a digital format and sends them to the computer.
- The plotter used to present the results of the data processing.
- A tape or CD/DVD drive is used for storing data or programs.
- A scanner or digitizer is required to convert the analogue data into digital format.

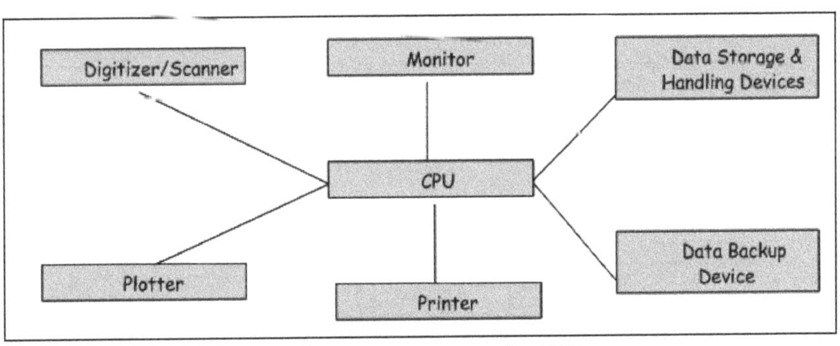

Hardware components for GIS.

The choice of hardware ranges from personal computers to multi user super computers. Computers should have essentially an efficient processor to run the software and sufficient memory to store data. The essential hardware elements for effective GIS operations include:

- The presence of a processor with sufficient power to run the software.
- Sufficient memory for the storage and backup of large volumes of data.
- A good quality, high resolution colour graphics screen or monitor.
- Data input and output devices, like keyboards, printers and plotters.

Software Components

GIS software components and sub-components have several functional elements to perform different operations. All GIS software fit this requirement; except in their user interface differences.

Table: GIS software functional elements.

Components	Sub-components
Data acquisition/Input	Digitizing
	Editing
Data processing and pre-processing	Topology building
	Projection transformation
	Format conversion
	Attribute assignment, etc.
Database management (storage and retrieval)	Data archival
	Hierarchical modelling
	Relational modelling
	Attribute query
	Object oriented database
Spatial manipulation and analysis	Measurement operations
	Buffering
	Overlay operations
	Connectivity operations
Product generation: Graphical output and visualization	Scale transformation
	Generalization
	Topographic maps
	Statistical maps

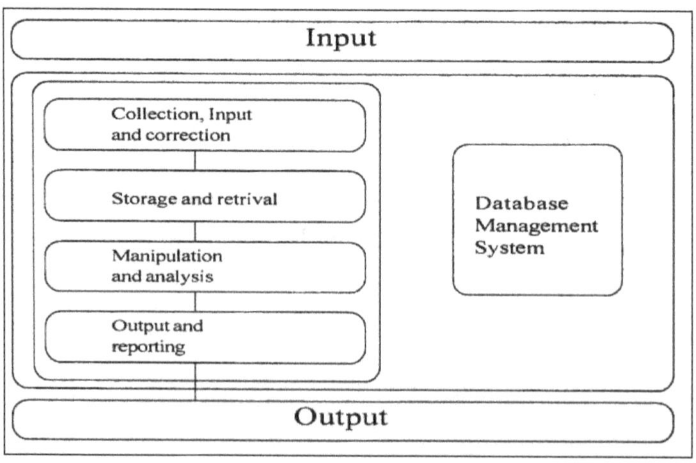

Workflow process of GIS - procedural perspective.

Figure also shows the procedural perspective workflow processes of GIS software to perform these functional elements though it is important to view these elements as a continuing process.

Data Input

Data input is the operation of encoding data and writing them on to the database for the GIS use. Data input involves data acquisition – i.e. identification and collection of the required data. It covers all aspects of transforming data captured from existing maps, field observations, and sensors into a compatible digital format. Data input can be performed in the following stages:

- Acquire or capture spatial data.
- Entering spatial and associated attributes of non-spatial data.
- Linking the spatial to the non-spatial data using unique identifiers.

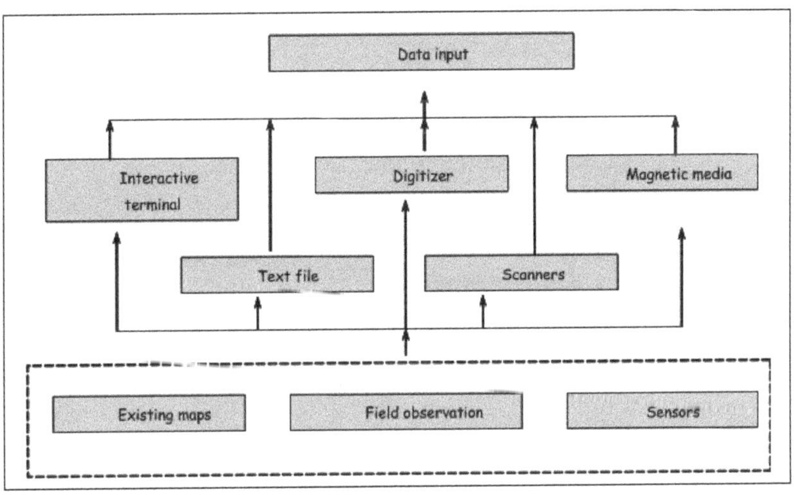

Data input for GIS.

Data Pre-Processing/Processing

Data input in a computer may involve several steps known as pre-processing. The procedure is used to convert a dataset into a format compatible for permanent storage within the GIS database and establishes a consistent system for data recording. Often, a large proportion of GIS data requires some kind of processing and data manipulation to get coordinated set of data layers. At this stage, the first digital map is constructed. The essential pre-processing procedures include:

- Format conversion such as geo-referencing with geometric correction and resampling, data generalization, and reduction (i.e. converting of GPS points into feature classes).
- Error detection and editing, edge matching and tiling.

- Merging of points into lines, and lines into polygons.
- Merging data storage and database management.
- Rectification/registration, interpolation, and photo-interpretation.

Data Storage and Database Management

The key idea to grasp about GIS software component is the geographic database management system. The data management functions necessary in any GIS facilitates the storage, organization and retrieval of data using a database management system (DBMS). It is a set of computer programs for organizing information, at the core of which will be a database. A database is a large, computerized collection of structured data. The computer program used to organise the database is a database management system (DBMS). It is a software package that allows the user to set up, use and maintain a database. All GIS software, regardless of vendor, consists of a DBMS capable of handling, organizing, and integrating of spatial data and attribute data.

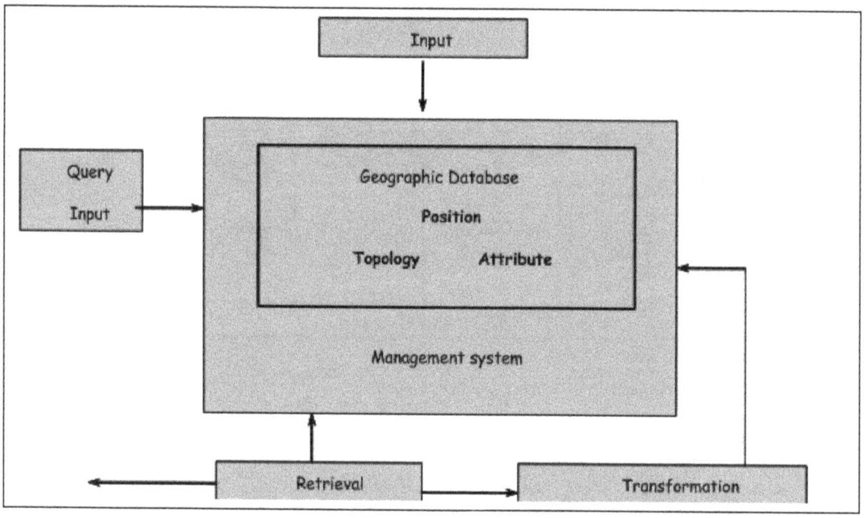

Data storage and database management system.

DBMS has to be structured and organized in such a way that they could be handled in the computer and how they are perceived by the users. It should indicate the way in which data about the position, linkages (topology), and attributes of geographical elements (such as points, lines, areas, and more complex entities representing objects on the earth's surface) are structured and organized. This function provides consistent method of data entry, update, deletion, and retrieval. An ideal GIS DBMS should provide support for multiple users and multiple databases to allow efficient updating minimize redundant information and data independence, security and integrity.

Data Analysis and Modelling

This functional element is the most important capability of GIS as far as the user is concerned and facilitates spatial analysis using spatial and non-spatial attributes. It involves working within databases to derive new information using several basic and advanced tools. This subsystem transforms spatial data, for example from one entity type (points, lines and areas) to another, and performs spatial analysis. Transformation may involve converting of raster's to vectors or vice versa. This is the distinguishing characteristics of GIS from other types of information systems. This subsystem allows us for an in-depth study of the topological and geometric properties of datasets. However, it is the most abused subsystem of a GIS due to lack of understanding the nature of spatial data contained in the subsystem. The operational classes of GIS can be grouped as:

- Retrieval, (re)-classification, and measurement functions.
- Overlay operations involve the combination of two or more datasets.
- Connectivity operations include contiguity, proximity, network, and spread operators.
- Neighborhood functions include search operations, topographic function, and interpolation.
- Modelling involves simplified representation and prediction of the reality (e.g. a land use map).

Data Output

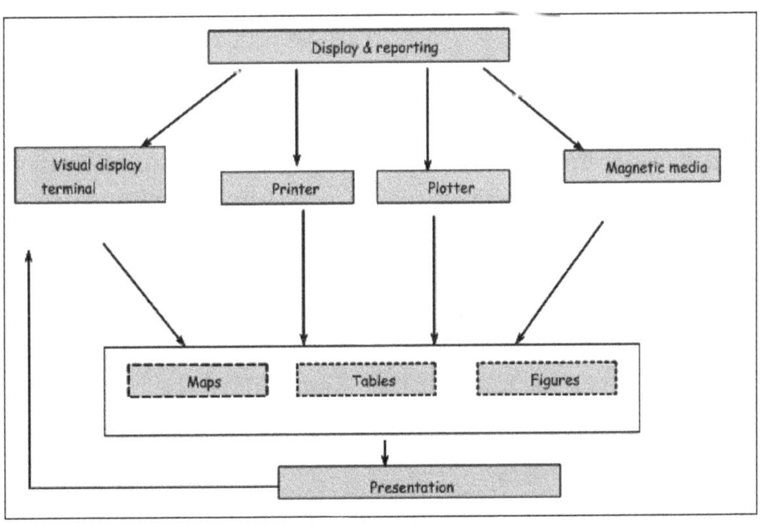

Data output for GIS.

This functional element concerns with the ways in which data are displayed and the results of analyses are reported to the users. These output products can be available in a variety of ways that includes statistical reports, maps, tables, figures, and graphics of various kinds. The generated product may range from the ephemeral image on a cathode ray tube through hard copy output drawn on printer or plotter to information recorded on magnetic media in digital form. There are several professional GIS software packages in the market such as ArcGIS, ILWIS, ERDAS, IDRISI, MAPINFO, GRASS, and the database systems of Oracle and dBase.

Data

Perhaps data is the most important component of a GIS. GIS can integrate spatial data with other existing data resources stored in a DBMS. All GIS software is designed to handle spatial data (also called geographical data). GIS technology utilizes two basic data types: attribute and spatial data.

- Attribute data describes the characteristics of spatial features often referred to as tabular data such as a name, number of stories, depth, or population.
- Spatial data refer to the real-world geographic objects such as streets, buildings, and countries, and their respective locations. They are represented in the form of vector data (points, lines and areas feature) or in the form of raster or image data (grid form).

GIS processes any data that has a spatial component. This information is quite diverse—it can be aerial photographs or satellite imagery, a collection of terrain contours, digital maps of the built environment, or legal records of land ownership. However, geographic data can reside in some unexpected places (i.e. any company that keeps a database of its customers has geographic data).

People or Expert

People are the key components of GIS, without which, nothing else would work or the GIS will be useless. Hence, there must always be people to plan, design, program, maintain it, supply it with data, interpret its results, implement and operate the system as well as make decisions based on the required output. People are the most complex component of GIS. This component is the most failure prone, and the only component capable of self-correcting; and able to repair other components. People may operate individually or within a group or team. Individuals may function as a GIS manager, analysts, technicians, or as a user. Almost all should have the basic knowledge needed to work with geographic data– knowledge of data sources, scale and accuracy, and software products.

GIS Manager	• Responsible to set the objectives for the GIS project, choose technologies and ensure the project to proceed smoothly. • May not be familiar with all the technical details of the project - focuses project-wide issues. • Responsible for accessing of GIS data for the GIS users.

GIS Analyst	• Responsible to ensure that the GIS solve the objectives set by the GIS Manager. • Must understand all the GIS functionality to program new applications. • Must understand and have experience on all the details of the GIS operation. • Should have troubleshooting skills--the ability to solve problems. • They need not be specialist in all disciplines of the GIS – the technician's role. • Responsible for accessing of GIS data for the GIS users.
GIS Technicians	• Perform data entry, data conversion, data correction, data analysis, and map production. • Not heavily involved in programming of the GIS. • They may make use of programs developed by the GIS Analyst. • Need to have the general understanding of how a GIS works.
GIS Users	• Make use of data stored in the GIS for scientific, technical, and policy analysis. • May request data in the form of maps. • They may work directly on the GIS, or they may use Internet to access GIS data. • Should understand what a GIS can do, and how to operate software.

Procedures or Method

A successful GIS operates according to a well-designed plan, business rules, procedures and methods unique to each organization. Sophisticated GIS technologies and new tools can only be used effectively if they are properly integrated into the entire business strategies and operations of any organization. It is simply not sufficient for an organization to purchase a computer with some GIS software, hire some enthusiastic individuals and expect instant success. In fact, the importance the application should not be undermined. From the application, a set of procedures will flow as a means by which organizations solve the problems inherent in the application. The procedures and software requirements will help to define a list of skills required to operate on the GIS functions, procedures, and make judgment. This component incorporates the following functions:

- Methods and procedures of data collection, encoding and storage in the GIS database.
- Methods and procedures of data quality assurance/verification to make sure that all data are correct, consistent, correct and up-to-date for analysis.
- Methods and procedures of data manipulation and query.

- Methods and procedures of data preparation and transformation of different sources.
- Techniques of data retrieval, operation, analysis, representation, presentation, and dissemination; and so on.

Functions of GIS

There are six major tasks carried out in any GIS software. They are data input, manipulation, management, query analysis and output or visualization (result).

Input

Two types of data are used in GIS. They are spatial and non-spatial data. The geographical location of any entity or feature is shown in the form of spatial data and attribute data are shown in non-spatial form which provide detailed information about spatial data. Therefore, attribute data should be logically attached with the concerned features. In GIS, there are five type of data entry system. They are as follows:

- Keyboard entry
- Coordinate geometry
- Manual digitizing
- Scanning
- Input of existing digital files

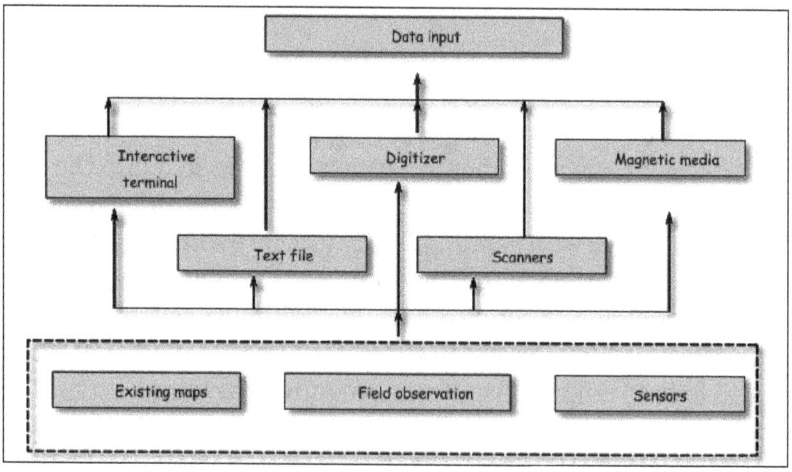

Manipulation

Manipulation procedure with data is applied to all components of geographic information system. These processes are carried out on the spatial data, attribute data and also on spatial relations among objects or structure elements of objects representation.

- Manipulation with Raster data: In the process of manipulation with raster attribute data the procedure carried out are:
    - Change of map projection.
    - Transformation of reference system.
    - Resampling of raster.
    - Conversion of representation.
    - Reclassification of raster values.
    - Assigning new values to raster.
    - Extraction of values from raster.
    - Creation and analysis of surfaces.
    - Mathematical and logical operations with raster.
- Manipulation of attribute data value of vector representation: In manipulation of attribute, data of vector following process are adopted:
    - Modification of topology.
    - Spatial splitting of vector representation.
    - Matching the edges of map sheets.
    - Change of raster cell size.
    - Generalization.

Management

There is enough storage space in computer files to store small GIS project data. However, it is difficult to store and manage large volume of data and large number of users. So in that case, data base management system is advised to use which helps in storage, management and organize large amount of data like tables, index, query and other processes in database. In DBMS, various types of models are there, but the relational model database management system is extremely useful in case of GIS use. Data are stored conceptually as a set of tables and every table will have the attribute data, which are related to a common entity in relational database model. With the help of this model, common fields of different tables are used to link with each other's. Relational database management system is very flexible in nature because of the simple architecture, so it is widely used in the GIS environment. With the help of Structured Query Language (SQL), the stored information of both spatial data and associated non-spatial attribute data can be retrieved. Query by using SQL depends upon the type of user interface. Besides this menu derived system is also used in retrieval of data.

Analysis

Analysis of geographic data is the most important task of GIS. The analysis process is also called as geographic or spatial analysis or geo-processing. Where to identify the patterns and trends and the future scenario the geographic properties of features are analysed. Therefore, there are various modern GIS tools for these types of analysis.

Visualization

Both spatial and tabular data can be presented in hard copy map, statistical summaries, modelling solutions and graphical representation in GIS platform. The end result in GIS is map and graphs, in which the result can be best represented. Best storing, communication and representation of geographic information is possible only through the maps and graphs, and GIS made it possible and easier. New and more innovative tools are there in GIS, which increase the art of visualization of the result or output to the users.

GIS Data Structure and Model

GIS does not store a map in any conventional sense, nor does it store a particular image or view of geographic area. Instead, a GIS stores the data from which we can draw a desired view to suit a particular purpose known as geographic data. There are two types of data in GIS; spatial data and non-spatial data (Attribute data). Non-spatial data include information about the features. For example, name of roads, schools, forests etc. population or census data for the region concerned etc. Non-spatial or attribute data is that qualifies the spatial data. It describes some aspects of the spatial data, not specified by its geometry alone. A geographical information system essentially integrates the above two types of data and allows user to derive new data for planning.

Spatial models are important in that way in which information is represented affects the type of analysis that can be performed and the type of graphic display that can be performed and the type of analysis that can be obtained. In GIS systems there is a major distinction between what are usually referred to as vector GIS and raster GIS. These two approaches to spatial data processing, often to be found in the same GIS package, reflect two different methods of spatial modelling: the former focusing on discrete objects that are to be described, and the latter concerned primarily with recording what is to be found at a predetermined set of locations that may be grid cells or points.

Spatial Information

Spatial characteristics of information can be broadly distinguished between those that describe where things are, using locations consisting of reference positions, spatial units and spatial relationships; those that describe the form of phenomena, using qualitative and quantitative descriptions of shape and structure; and those that describe associations and interactions between different phenomena. All geographical data can be reduced to three basic geographical phenomenon can in principle be represented by a point, line or area plus a label saying what it is. So an oil well could be represented by a point entity consisting of a XY coordinate; a road could be represented by a series of XY coordinates; a floodplain could be represented by an area entity covering a set of XY co-ordinates plus the label 'floodplain'. The labels could be the actual names as given here, or they could be special symbols.

The essential features of any data storage system are that they should be able to allow data to be accessed-and cross-referenced quickly. There are several ways of achieving this, some of which are more efficient than others. Unfortunately, there seems to be no one 'best' method that can be used for all situations. This explains in part the massive investment in labour and money in effective database management systems, which are the computer programs that control data input, output, storage, and retrieval from a digital database.

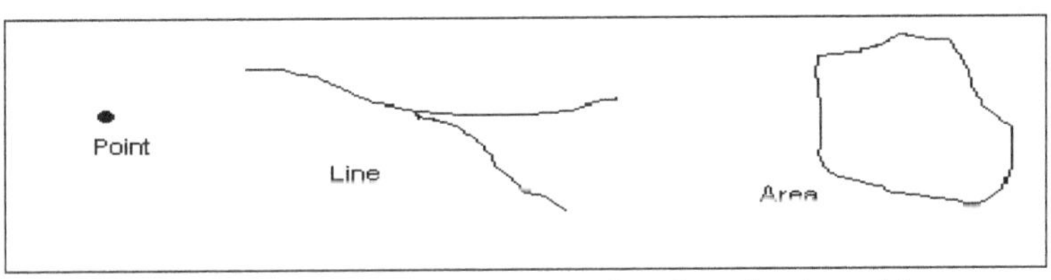

Real world phenomena represented as three basic entities.

Layers and Coverage

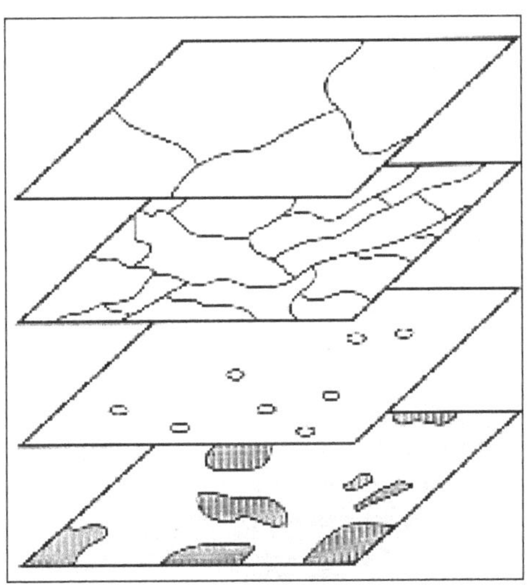

The Common requirement to access data on the basis of one or more classes has resulted in several GIS employing organizational schemes, in which all data of a particular level of classification, such as roads, rivers or vegetation types, are grouped into layers or coverage. GIS organize spatial data into layers. Typical layers represent information belonging to particular classes. The layers can be combined with each other in various ways to create new layers that are a function of the individual ones. Any layer does not contain any areal regions that are overlapping; therefore it is possible for each region to have multiple attributes corresponding to multiple perspectives on the meaning of that region.

Data Model

In order to represent the spatial information and their attributes, a data model – a set of logical definitions or rules for characterizing the geographical data is adopted. The data model represents the linkages between the real world domain of geographical data and the computer and GIS representation of these features. As a result, the data model, not only helps in organizing the real-world geographical features into a systematic storage/retrieval mechanism, but also helps in capturing the user's perception of these features. The model:

- Structures the data to be amenable to computer storage/retrieval and manipulation. The data structure is the core of the model and it is based upon this that features of real world are represented. The ability of the data structure to totally represent the real world determines the success of the model.
- Abstracts the real world into properties, which is perceived by a specific application. For example, a Land use map is perceived to be made up of different classes with symbols and legends. The district information is perceived to be made up of district maps and different attribute tables.
- Helps organize a systematic file structure, which is the internal organization of real world data in a computer.

Conceptual Models of Spatial Information

There are different models, which have influenced the way in which data are organized and processed within GIS. They are based respectively on objects, networks and fields.

Object-Based Model

Object-based spatial models emphasize individual phenomena that are to be studied in isolation or in terms of their relationships with other phenomena. Any phenomena however big or small may be designated as an object, provided that it can be separated conceptually from neighboring phenomena. Objects may be composed from other objects and they may have specific relationships with other separate objects. An object-based view is appropriate, though not confined, to phenomena that have well defined boundaries. Hence it is suited to human made phenomena, such as buildings, roads, utilities and administrative regions. Some natural phenomena, such as lakes, rivers, islands and forests, are often represented in object based models because they need to be treated as discrete phenomena for some purposes.

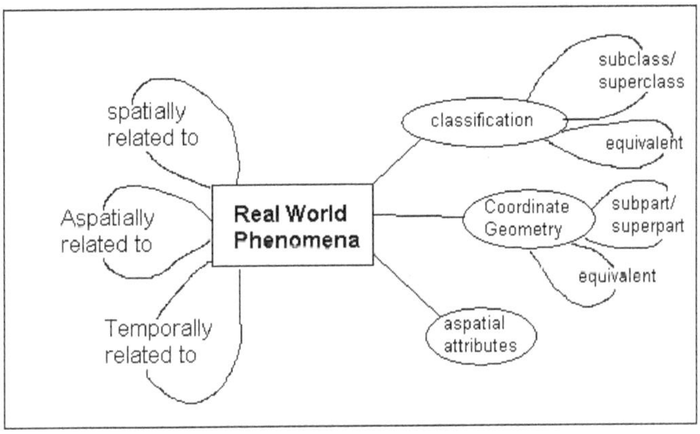

The object based conceptual view.

Networks Model

Network based spatial models share some aspects of the object based model in that they often deal with discrete phenomena, but the essential characteristic is the need to consider interactions between multiple objects, often along discrete paths or routes that connect them. The exact shape of the phenomena may not be of much importance. What is more important is some measure of distance, or the impedance, between specified phenomena. Typical examples of applications for which a network model is appropriate are studies of traffic on Road Sea and air route; and analysis of flow of water, gas and electricity through pipelines and cables.

Fields Model

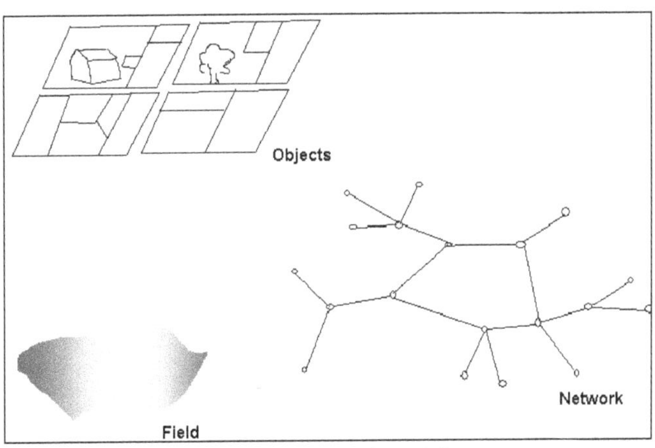

Conceptual models of spatial Data.

The field based view is appropriate for modelling phenomena that are regarded as continuously variable across some region of space. Examples of phenomena that may be treated as fields include the concentration of pollutants in the atmosphere, temperature of the ground surface, the moisture levels in soil, and the speed and direction of flow in bodies of air and water. Fields may represent either two or three spatial dimensions, depending on the application. A field based spatial model is often adopted when the data to be modelled are not known in sufficient detail to provide precise boundaries, even though at some resolution they could be said to exist.

Spatial Data Models for GIS: Representation in Computer

When Geographical data are entered into a computer the user will be most at ease if the geographical information system can accept the phenomenological data structures that he has always been accustomed to using. But computers are not organized like human minds and must be programmed to represent phenomenological structures appropriately. Moreover, the way the geographical data are visualized by the user is frequently, not the most efficient way to structure the computer database. Finally, the data have to be written and stored on magnetic devices that need to be addressed in a specific way.

Geographical information systems provide methods for representing spatial data that allow the user to adopt conceptual models resembling to a large extent the three classes of model as discussed above. There are two broad categories of spatial data models. These are vector data model and raster data models. The data base concept is central to a GIS and is the main difference between a GIS and drafting or computer mapping systems, which can produce only good graphic output. All contemporary Geographic Information system incorporates a data base management system. Data base systems provide the means of storing a wide range of geographic information and updating it without the need to rewrite program. In GIS, the spatial data models handle where the features are and Non-spatial data models or Data base management system handle the feature description and how each feature is related to other.

Two approaches or models have been widely adopted for representing the spatial data within GIS; The cartographic map model and the geo-relational model. Each of these approaches is based on a specific spatial data model. The Composite Map Model is usually based on a tessellated (raster) representation of space and the Geo-relational model is usually associated with a vector representation of space.

Vector data model represents phenomena in terms of the spatial primitives, or components, consisting of point, line, polygon, surfaces and volumes. Raster data model represents phenomena as occupying the cells of a predefined, grid shaped tessellation.

Raster Data

Raster data presented as a matrix or array of pixels (Picture Element). Pixel is the smallest unit of picture. It is stored spatial information in grid cells organized and accessed as rows and columns. The information of geographic features stored within the cells contains a number called Digital Number (DN) value. DN values of geographic features are dependent on feature's characteristics and reflectance value over the surface of the earth. Each geographic feature has own DN value arranged in regular grid. The spatial resolution of raster data is dependent on cell or pixel size. Spatial resolution of raster data increased with decreasing pixel size. Spatial details of raster data controlled by resolution. There are many sources of raster data such as satellite imagery, aerial photographs, scanned maps and other sources.

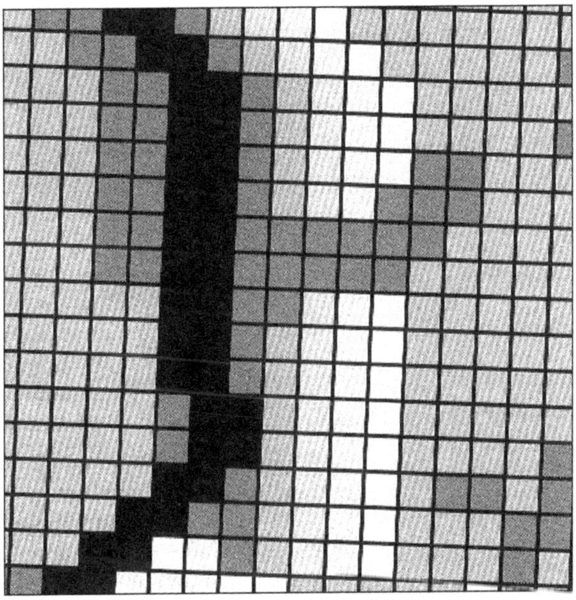

Properties of Raster Data

- Raster data is a set of cells located by coordinate is used.
- Each cell is independently addressed with the value of an attribute.
- Each cell contains a single value and every location corresponds to a cell.
- One set of cell and associated value is a LAYER.
- The linear dimension of each cell defines the spatial resolution of the data.
- The minimum mapping unit on 1:50000 scales are 3mm x 3mm or 2.5 hectares.
- Raster data models require a huge volume of data to be stored, fitness of data be limited by cell size.

Raster data is divided into following two categories:

- Discrete data: It is also called categorical, thematic, or discontinuous data. It is used to present discrete features in both raster and vector data models. The boundary of discrete features has known. For example, a forest is a discrete feature within the surrounding landscape. A discrete feature is represented using same value neighbouring cells in raster data model. Roads, water bodies and built-up area are the examples of discrete features.

- Continuous data: It is also called surface data or non-discrete data. Continuous data is divided into two types based on features they represent. First, to represent features using continuous data the value at each cell location measured by fixed registration point. For example height measured from ground surface as the fixed point. The another type of continuous data shows feature characterized by the way they move and represent features that gradually vary as they move across the surface from a source such as liquid and gas movement. It is impossible to measure every cell location of all continuous data because they are derived from discontinuous data. Interpolation method is used to obtain the continuous surface, which is based on the features characteristics.

There are various methods for encoding raster data from scratch. Few models are as follows:

- Cell-by-cell raster encoding: This method encodes a raster by creating records for each cell value arranged in row and column. According to this method a big spreadsheet consists of cells and each cell represents a pixel in the raster data. This method is also known as exhaustive enumeration.

- Run-length raster encoding: In this method cell values encodes in runs of similar valued pixels and can result in highly compressed image data. This method is useful when large groups of neighboring pixels have similar values or less useful when neighboring pixel values vary widely.

- Quad-tree raster encoding: In this method raster data is divided into hierarchy of quadrants and sub-divided based on similar valued pixels. When a quadrant is made completely from the similar value cells the division of the raster data is stops. A quadrant that cannot be sub-divided is known as leaf node.

Vector Data

Vector Spatial Data Model

In the second and more intensely developed approach to information integration, attribute information is associated with point, line and polygons – as spatial entities that describe features occurring in the real world. Thus, for example, a point feature such as a city may have associated with it such items of information as its total population, number of houses, and number of schools and so on. Similarly, a linear feature such as a river might have associated with it such information as name, mean discharge etc. A polygonal feature such as land use category might be linked to information describing its use, past land use, its soil type and so forth.

The vector representation of an object is an attempt to represent the object as exactly as possible. The co-ordinate space is assumed to be continuous, not quantised as with the raster space, allowing all positions, lengths, and dimensions to be defined precisely.

Vector data models treat phenomena as sets of primitives and compose spatial entities. In 2D models the primitive entities are points, lines and areas. The vector data structure represents each geographical feature by a set of coordinates. Vectors as x,y coordinates define points, lines and polygons. The basic premise of the vector based structuring is to define a 2-dimensional space where coordinates on the two-axes represent features. Generally, representing points and lines is straightforward – points are characterized by a x,y coordinate pair and line by a set of x,y coordinate pairs with a specific beginning and ending vector. However, representing polygons in vector storage poses a challenge. The three vector structures used in Geographical Information Systems (GIS) for the storage of points lines, and areas are:

- Point entities: Point entities can be considered to embrace all geographical and graphical entities that are positioned by a single XY co-ordinate pair. Besides the XY coordinates, other data must be stored to indicate what kind of 'point' it is, and the other information associated with it.
- Line entities: Line entities can be defined as all linear features built up of straight line segments made up of two or more co-ordinates. The simplest line required the storage of a begin point and an end point (two XY coordinate pairs) plus a possible record indicating the display symbol to be used. An 'arc', a 'chain' or a 'string' is a set of n XY co-ordinates pairs describing a continuous complex line. The shorter the line segments, and the larger the number of XY co-ordinates pairs, the closer the chain will approximate a complex curve. Data storage space can be saved at the expense of processing time by storing a number that indicates that the display driver routine should fit a mathematical interpolation function to the stored co-ordinates when the line data are sent to the display device. As with 'point' and simple line, chains can be stored with data records indicating the type of display line symbols to be used.
- Area entities: There are several ways of vector structures possible for structuring polygons. The simplest way to represent a polygon is the spagetti representation, which is nothing, but an extension of the simple chain, i.e. to represent each polygon as a set of XY co-ordinates on the boundary i.e. the polygons are discretised to the concept of line representation and are characterized by a set of xy coordinate pairs, but have the same vector as the beginning and ending vector that is representing a self-closing line as a polygon. The name of the symbols are used to tell the user what each polygon in ard then held as a set of simple text entities.

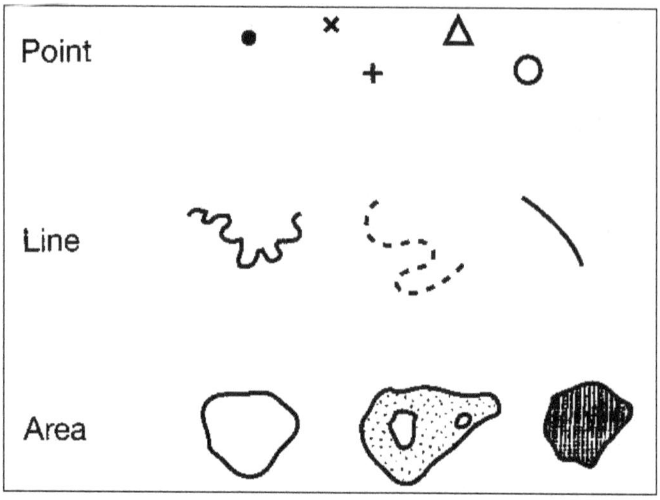

Geometric primitives categorized as point, line and area.

Topology

An important aspect of vector-based models is that they enable individual components to be isolated for the purpose of carrying out measurements of, for example, area and length, and for determining the spatial relationships between the components. Spatial relationships of connectivity and adjacency are examples of topological relationships and a GIS spatial model in which these relationships are explicitly recorded is described as topologically structured. In a fully topologically structured data set, wherever lines or areas cross each other, nodes will be created at the intersections and new areal subdivisions defined.

Vector-based spatial data models that are topologically structured are often described in a terminology of topological objects or primitives, which are classified in terms of topological dimensions, where an n-cell topological object is of n dimensionality. Two-dimensional topological objects consist of polygons (faces or areas), arcs and nodes. For areal information, adjacency between the areas can be recorded in terms of feature codes associated with the left and right sides of arcs. The expressions 'left' and 'right' are given meaning in this context by specifying the direction of the arc in terms of a 'from node' and a 'to node'.

The composition of each polygon can be defined by listing the component arcs, including a negative sign where necessary to ensure consistency in arc direction. In order to distinguish between external and internal boundaries, a convention of clockwise for the former and anticlockwise for the latter (or vice versa), can be adopted. For the purpose of network analysis, each node may be associated with a list of the arcs, which it bounds. The list of arcs connected to each node will generally be in a predetermined order, namely clockwise or anticlockwise.

Topological structure is important in keeping track of the components of complex objects and in determining the spatial relationships of connectivity and adjacency between recorded phenomena. Thus if two lines cross each other they will share a common node. If two areas are adjacent to each other, such as two neighboring counties, they will share a common boundary arc. If the boundary of a county coincides with the path of a river they might also share the same arc. The inclusion of one area in another, such as a specific type of forest within a county, will result in their sharing common polygons.

The presence of these various spatial relationships can be determined by relatively simple comparisons of the identifiers of their topological components, rather than requiring possibly computationally demanding geometric calculations based on coordinates. It may also be noted that because shared spatial objects are only stored once, though perhaps referenced many times, storage space is saved by avoiding duplication of the same geometric data. This in turn assists in the maintenance of the integrity of the database by avoiding the possibility of two different versions of the same geometric components.

The Choice Between Raster and Vector

Only a few years ago, the conventional wisdom was that raster and vector data structures were irreconcilable alternatives. They were then irreconcilable because raster methods were required huge computer memories to store and process image at the level of spatial resolution obtained by vector structures. Certain kinds of data manipulation, such as polygon intersection or spatial averaging presented enormous technical problems with the choice of raster methods that allowed easy spatial analysis but resulted in ugly maps, or vector methods that could provide database of manageable size and elegant graphics but in which spatial analysis was extremely difficult.

Vector Methods

Advantages:

- Good representation of phenomenological data structure.
- Compact data structure.
- Topology can be completely described with network linkages.
- Accurate graphics.
- Retrieval, updating and generalization of graphics and attributes are possible.

Disadvantages:

- Complex data structures.
- Combination of several vector polygon maps or polygon and raster maps through overlay creates difficulties.

- Simulation is difficult because each unit has a different topological from.
- Display and plotting can be expensive, particularly for high quality, colour and crosshatching.
- The technology is expensive, particularly for the more sophisticated software and hardware.
- Spatial analysis and filtering within polygons are impossible.

Raster Methods

Advantages:

- Simple data structures.
- The overlay and combination of mapped data with remotely sensed data is easy.
- Various kinds of spatial analysis are easy.
- Simulation is easy because each spatial unit has the same size and shape.
- The technology is cheap and is being energetically develop.

Disadvantages:

- Volumes of graphic data.
- The use of large cells to reduce data volumes means that phenomenological. Recognizable structures can be lost and there can be a serious loss of information.
- Crude raster maps are considerably less beautiful than maps drawn with fine lines.
- Network linkages are difficult to establish.
- Projection transformation is time consuming unless spatial algorithms or hardware are used.

The problem of raster or vector disappears once it is realized that both are valid methods for representing spatial data, and that both structure are interconvertible. Conversion from vector to raster is the simplest and there are many well know algorithms. Vector to raster conversions are now performed automatically in many display screens by inbuilt microprocessors. The reverse operation raster to vector, is also well understood (Pavlidis lists four algorithms for thinning bands of pixel to lines), but it is a much more complex operation that is complicated by the need to reduce the number of co-ordinates in the resulting lines by a process known as weeding.

The Use of Raster and Vector Methods

- Use Vector data structure for data archiving phenomenological structured data (e.g. soil areas, land use units, etc).
- Use Vector methods for network analyses, such as for telephone networks, or transport network analysis.
- Use Vector data structure and Vector display methods for the highest quality line drawing.
- Use Raster methods for quick and shear map over lay, map combination and spatial analysis.
- Use Raster methods for simulation and modelling when it is necessary to work with surfaces.
- Use Raster and Vector in combination for plotting high quality lines in combination with efficient area filling in colour. The lines can be held in Vector format and the raster filling in compact Raster structures such as run length codes or quadtrees.
- Preferably use compact Vector data structure for digital terrain models, but don't neglect altitude matrices.
- Use Raster - Vector and Vector - Raster algorithms to convert data to the most suitable form for a given analysis or manipulation.
- Remember that Display systems can operate either in Raster or Vector modes independent of the Data Structures that are used to store and manipulate the data.

GIS Database Creation

In GIS getting the spatial data is not an easy and cheap task. Most costlier and time consuming phase of GIS is data capturing. For instance, spatial data capturing constitute about 60 to 80 per cent of total cost of fully operational GIS based projects.

GIS Data Source

Spatial data are collected from mainly following two sources. They are:

- Primary source: In primary data, source spatial data is captured directly into GIS environment. For instance, capturing of data from Global positioning system (GPS) or through satellite imagery from remote sensing satellite. Attribute data is collected through surveying and also through GPS.

- Secondary source: Capturing of data from Secondary source is easier task and more common in the field of History. In this source hard copy map is converted into digital form. For conversion of hard copy map into digital form generally following two techniques are used. First one is to produce raster format data or digital data, scanning of hard copy map is carried out. In second process hard copy map is digitised, where point, line and polygon features are traced. There are also two ways to doing this. First on is directly tracing of hard copy map into digital format and second one is digitization of scanned map. Vector data is produced in these processes. Attribute data are collected by already published sources like census, district abstracts, research papers, articles and books etc.

Methods of GIS Data Creation

The power of a GIS is in the data analysis. Without data, all the bells and whistles of GIS are just that. The effort to bring stable, accurate data is an enormous one for any GIS. Data development and maintenance is the most costly, labor intensive part of developing a GIS. There are several ways in which to bring spatial data into a GIS.

Manual Digitizing

This is traditionally the most common way to convert paper-based sources of spatial information (e.g. maps) to digital data. The paper map is attached by tape to a digitizing table (or tablet as the smaller digitizers are known). Usually between 4-6 initial points of which the coordinates are known are logged. Optimally these points are such locations as the intersections of graticule lines. In the absence of an overlying grid system, points are taken from identifiable locations such as street intersections or landmarks.

The data is then digitized by tracing the features of interest with a mouse like hand held device called a puck. Once all the features are traced the newly acquired data is transformed from table units (the coordinates of the digitizing table) to real world units using and algorithm. This algorithm takes the known table coordinates of the initial points and warps the data to match the real world coordinates assigned to those points. To error in the adjustment from the table units to real world coordinates is called the RMS error. Results are reported as root-mean-square (RMS) error and average error. The RMS value reflects the range of the error; the precision of the digitized data. Factors contributing to this error can be human error, shrinkage or physical alteration of the paper map and projection differences.

Heads up Digitizing

With the proliferation of low cost sources of digital imagery and large format scanners, heads up digitizing is becoming a popular method of digital conversion. Also known as on-screen digitizing, this method involves digitizing directly on top of an orthorectified image such as a satellite image or an aerial photograph. The features of interest are traced from the image. The benefit of this over manual digitizing is that no transformation is needed to convert the data into the needed projection. In addition, the level of accuracy of the derived dataset is taken from the initial accuracy of the digital image. Heads-up digitizing is also utilized in extracting data from scanned and referenced maps.

Coordinate Geometry (COGO)

Coordinate geometry is a keyboard-based method of spatial data entry. This method is most commonly used to enter cadastral or land record data. This method is highly precise as entering the actual survey measurements of the property lines creates the database. Distances and bearings are entered into the GIS from the original surveyor plats. The GIS software builds the digital vector file from these values.

Geocoding

Geocoding is another keyboard-based method. Geocoding uses addresses from a flat file (such as a.dbf file, MS Access database or excel spreadsheet) to create x,y coordinate locations interpolated from a geocodable spatial database. These spatial databases are most commonly street centerline files but can be other types. The resultant geocoded database is a point file.

Global Positioning Systems (GPS)

GPS is a way to gather accurate linear and point location data. Originally devised in the 1970s by the Department of Defense for military purposes, the current GPS consists of 28 satellites that orbit the earth, transmitting navigational signals. Through interpolation, these signals received by a data logger can pinpoint the holder's location. Depending on the unit, the locational accuracy can reach to the millimeter. Combined with attribute data entered at the time of collection, GPS is a rapid and acccurate method of data collection.

Image Processing

Geodatasets can be derived from digital imagery. Most commonly satellite imagery is utilized in a process called supervised classification in which a user selected a sampling of pixels for which the user knows the type (vegetation species, land use, etc). Using a classification algorithm, remote sensing software such as ERDAS or ENVI classifies a digital image into these named categories based on the sample pixels. In contrast to the other methods discussed, supervised classification results in a raster dataset.

Types of Digitizing Errors in GIS

Since most common methods of digitizing involve the interpretation of geographic features via the human hand, there are several types of errors that can occur during the course of capturing the data. The type of error that occurs when the feature is not captured properly is called a positional error, as opposed to attribute errors where information about the feature capture is inaccurate or false.

During the digitizing process, vectors are connected to other lines by a node, which marks the point of intersection. Vertices are defining points along the shape of an unbroken line. All lines have a starting point known as a starting node and an ending node. If the line is not a straight line, then any bends and curves on that line are defined by vertices (vertex for a singular bend). Any intersection of two lines is denoted by node at the point of the intersection.

Dangles or Dangling Nodes

Dangles or dangling nodes are lines that are not connected but should be. With dangling nodes, gaps occur in the line work where the two lines should be connected. Dangling nodes also occur when a digitized polygon doesn't connect back to itself, leaving a gap where the two end nodes should have connected, creating what is called an open polygon.

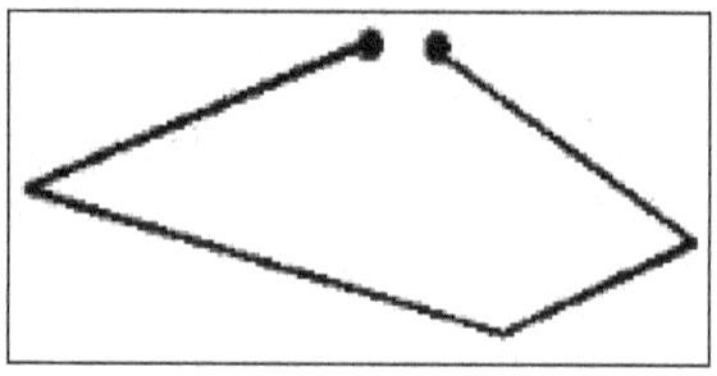

An open polygon caused by the endpoints not snapping together.

Switchbacks, Knots and Loops

These types of errors are introduced when the digitizer has an unsteady hand and moves the cursor or puck in such a way that the line being digitized ends up with extra vertices and/or nodes. In the case of switchbacks, extra vertices are introduced and the line ends up with a bend in it. With knots and loops, the line folds back onto it, creating small polygon like geometry known as weird polygons.

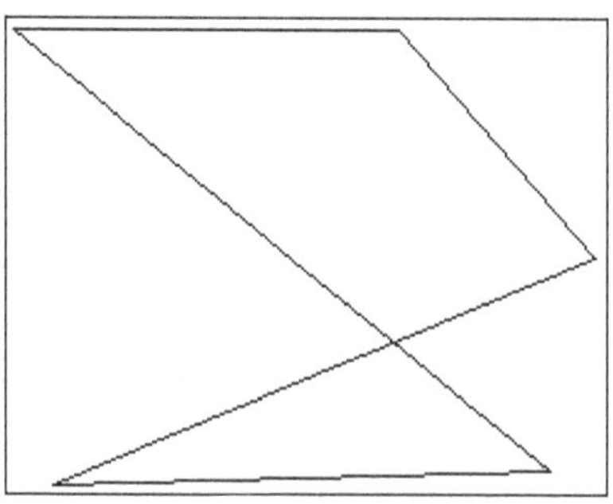

Example of a weird polygon where the line folds back on itself.

Overshoots and Undershoots

Similar to dangles, overshoots and undershoots happen when the line digitized doesn't connect properly with the neighboring line it should intersect with. During digitization a snap tolerance is set by the digitizer. The snap tolerance or snap distance is the measurement of the diameter extending from the point of the cursor. Any nodes of neighboring lines that fall within the circle of the snap tolerance will result in the end points of the line being digitized automatically snapping to the nearest node. Undershoots and overshoots occur when the snap distance is either not set or is set too low for the scale being digitized. Conversely, if the snap distance is set too high and the line endpoint snaps to the wrong node. In a few cases, undershoots and overshoots are not actually errors. One instance would be the presence of cul-de-sacs (i.e. dead ends) within a road GIS database.

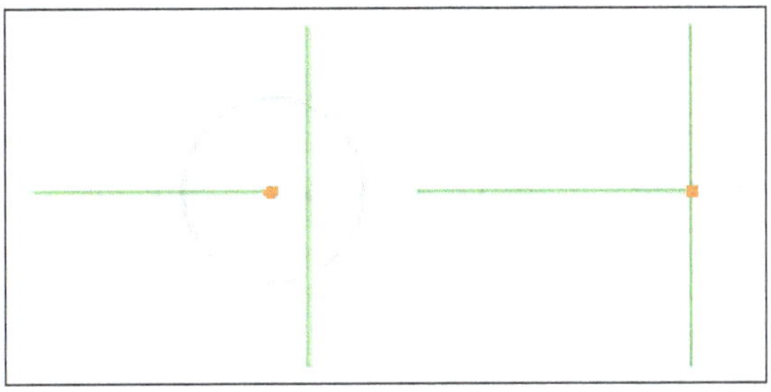

The circle represents the area of the snap tolerance. The line being digitized will automatically snap to the nearest nodes within the snap tolerance area.

Slivers

Slivers are gaps in a digitized polygon layer where the adjoining polygons have gaps between them. Again, setting the proper parameters for snap tolerance is critical for ensuring that the edges of adjoining polygons snap together to eliminate those gaps. Where the two adjacent polygons overlap in error, the area where the two polygons overlap is called a sliver.

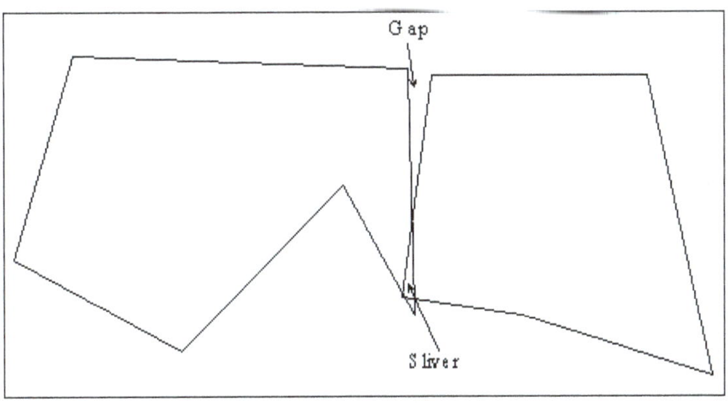

Gap and Sliver Errors in Digitized Polygons.

2

Principles and Concepts of Remote Sensing

The process of scanning and recording the physical properties of an object or phenomena without making physical contact with the object is known as remote sensing. This chapter discusses some of the important topics related to remote sensing like its platforms and sensors and remote sensing in solar reflection region. These topics are crucial for a complete understanding of the subject.

Remote Sensing

Remote sensing is a method of collecting information about any ground object under investigation from a distance without being in contact. The most accepted definition was given by the American Society for Photogrammetry and Remote Sensing in 1988. Accordingly, Remote Sensing can be defined as "The art, science and technology of obtaining reliable information about physical objects and the environment, through the process of recording, measuring and interpreting imagery and digital representations of energy patterns derived from non-contact sensor systems."

Remote sensing is one of a suite of geospatial technologies that are having a growing impact in a wide variety of areas from commerce to science to public policy. The field of remote sensing evolved from the interpretation of aerial photographs to the analysis of satellite imagery, and from local area studies to global analyses, with advances in sensor system technologies and digital computing. Today remote sensor systems can provide data from energy emitted, reflected, and/or transmitted from all parts of the electromagnetic spectrum. Examples of applications of these data include population and demography studies, study of archaeological sites, energy studies using hydrological models, urban planning, environmental treaty enforcement, and agricultural yields.

Although remote sensing techniques have primarily been viewed as a means for gathering data that are then interpreted by the user, they are increasingly serving other roles in scientific and applied research. Remotely sensed data on natural and anthropogenic features such as vegetation cover, land use, topography, and hydrography now serve as input to a range of simulation models, including hydrologic, climatic, ecological, and economic models. Classified images of land use and land cover are combined with surveys of demographic and socioeconomic variables to develop models that allow scientists to better understand processes such as deforestation and land use conversion.

Remotely sensed imagery and the products derived from analysis of imagery are also important sources of data for geographic information systems (GIS). In fact, most comprehensive image analysis software packages now include GIS functions for change detection overlays, local spatial analysis techniques, conversions between raster (i.e., pixel-based grids) and vector (i.e., points, lines, and polygons defined and displayed on the basis of two-dimensional Cartesian coordinate pairs) data structures, and other not strictly image-related processes. GIS software packages by necessity work with raster data and images in a number of formats, and they increasingly include analysis functions that were previously only found in specialized image analysis packages.

Compared to field-based sampling, remote sensing cannot provide measures of human and environmental phenomena, such as water quality, vegetation composition, soil properties, or plant health, with the same amount of detail. However, when coupled with field surveying, remote sensing offers the ability to view and map large areas of the earth's surface at multiple times and to obtain information for areas that would otherwise be difficult or impossible to sample due to physical or financial constraints. Remote sensing has thus become a valuable tool in research and applications in a wide range of disciplines, such as engineering, geology, geography, urban planning, forestry, and agriculture. Furthermore, the Internet has increased the availability and dissemination of remote sensing products, and decreasing costs coupled with continuous improvements in spatial, spectral, radiometric, and temporal resolutions are making remote sensed data accessible to a broader range of end users and expanding the role of remote sensing in society.

Passive Versus Active Remote Sensing

Remote-sensing instruments are often categorized as having either active or passive sensors. An active sensor generates its own signal which is subsequently measured when reflected back by the Earth's surface. A passive sensor measures solar energy that is either reflected or emitted from features on the Earth's surface. Table below lists a number of different active and passive instruments mounted on satellite platforms.

Table: Active and passive satellite-based remote-sensing instruments.

Sensor name	Type	Wavelength range	Resolution (m)
IKONOS	Optical	450–900 nm	1–4
SPOT5a	Optical	500–1750 nm	2.5–10
IRS-P6 – LISS-4	Optical	520–860 nm	5.8
ALOS AVNIR-2	Optical	420–500 nm	10
ASTER	Optical	520–11 650 nm	15–60
Landsat ETM+	Optical	450–2350 nm	15–30
MODIS	Optical	459–14 385 nm	250–1000
AVHRR	Optical	580–12 500 nm	1000
ENVISAT	Radar	5.7 cm (C band)	25

RADARSAT-1	Radar	5.7 cm (C band)	10–100
RADARSAT-2	Radar	5.6 cm (C band)	3–100
ALOS – PALSAR	Radar	23.5 cm (L band)	10–100

Although most passive sensors operate in the visible and infrared portions of the EMS, there are also some passive microwave sensors in use that measure a number of parameters such as wind speed, atmospheric and sea surface temperature, soil moisture, rainfall, and atmospheric water vapor.

An advantage of passive sensors is that most rely on the Sun's energy to illuminate the target and therefore do not need their own energy source so in general they are simpler instruments. A limitation for most passive optical sensors is that they require daylight to operate, although there are some sensors that record night time lights and clouds at night and others that record energy emitted from the Earth's surface. Since most of these sensors operate in the visible and infrared wavelengths, they are adversely affected by weather and cloud cover. Lastly, since sunlight is primarily reflected from the top of a feature, such as a forest, it is not possible to 'see' under a canopy to measure vegetation structure. To obtain this kind of information it is necessary to use active sensors.

Active sensors, such as radar and lidar emit their own energy to illuminate a target and are comprised of a signal generator and receiver. They measure the strength of the returned signal and the time delay between when the instrument emits the energy and when it receives the returned pulse. These two types of information are used to describe vegetation structure. Radar is an acronym for 'radio detection and ranging'. Radar systems operate in the long-wavelength microwave portion of the EMS and thus are largely unaffected by clouds and rain. They can be considered all-weather systems. Lidar is an acronym for 'light detection and ranging' and these systems use lasers that emit light in the visible and near-infrared portions of the EMS. In a lidar system a single light pulse can reflect off of several features in vertical space such as different layers in a forest. A single emitted pulse will result in a wave or series of returned pulses that are recorded by the detector. These return pulses can be recorded as a wave (full-waveform lidar) or in discrete pieces that correspond to the peaks in the returned signal. A number of different types of lidar systems have been developed but most provide the capability to record the first and last return of the light pulse. The returns correspond to the top of an object (i.e., top of a tree canopy) and the base substrate that the object is resting on (i.e., the ground). This is ideal for measuring the height of features such as trees or buildings.

Radar systems behave differently from optical systems with respect to how they interact with materials. The signal from most radar systems can penetrate well into a forest canopy, and radar systems with especially long wavelengths (e.g., P-band systems) can even penetrate dry ground. Although lidar is probably best known for its capability of acquiring digital elevation data which can be used to describe topography, these systems show a lot of promise to be able to directly measure vegetation structure characteristics. Lidar can make direct measurements of vegetation structure and provide vertical information about structure that is largely missing in data collected using passive remote-sensing instruments which only record light reflected from the top of a canopy.

Advantages of Remote Sensing Technology

- Large area coverage: Remote sensing allows coverage of very large areas which enables regional surveys on a variety of themes and identification of extremely large features.
- Remote sensing allows repetitive coverage which comes in handy when collecting data on dynamic themes such as water, agricultural fields and so on.
- Remote sensing allows for easy collection of data over a variety of scales and resolutions.
- A single image captured through remote sensing can be analyzed and interpreted for use in various applications and purposes. There is no limitation on the extent of information that can be gathered from a single remotely sensed image.
- Remotely sensed data can easily be processed and analyzed fast using a computer and the data utilized for various purposes.
- Remote sensing is unobstructive especially if the sensor is passively recording the electromagnetic energy reflected from or emitted by the phenomena of interest. This means that passive remote sensing does not disturb the object or the area of interest.
- Data collected through remote sensing is analyzed at the laboratory which minimizes the work that needs to be done on the field.
- Remote sensing allows for map revision at a small to medium scale which makes it a bit cheaper and faster.
- Color composite can be obtained or produced from three separate band images which ensure the details of the area are far much more defined than when only a single band image or aerial photograph is being reproduced.
- It is easier to locate floods or forest fire that has spread over a large region which makes it easier to plan a rescue mission easily and fast.
- Remote sensing is a relatively cheap and constructive method reconstructing a base map in the absence of detailed land survey methods.

Disadvantages of Remote Sensing

- Remote sensing is a fairly expensive method of analysis especially when measuring or analyzing smaller areas.
- Remote sensing requires a special kind of training to analyze the images. It is

therefore expensive in the long run to use remote sensing technology since extra training must be accorded to the users of the technology.

- It is expensive to analyze repetitive photographs if there is need to analyze different aspects of the photography features.
- It is humans who select what sensor needs to be used to collect the data, specify the resolution of the data and calibration of the sensor, select the platform that will carry the sensor and determine when the data will be collected. Because of this, it is easier to introduce human error in this kind of analysis.
- Powerful active remote sensing systems such as radars that emit their own electromagnetic radiation can be intrusive and affect the phenomenon being investigated.
- The instruments used in remote sensing may sometimes be un-calibrated which may lead to un-calibrated remote sensing data.
- Sometimes different phenomena being analyzed may look the same during measurement which may lead to classification error.
- The image being analyzed may sometimes be interfered by other phenomena that are not being measured and this should also be accounted for during analysis.
- Remote sensing technology is sometimes oversold to the point where it feels like it is a panacea that will provide all the solution and information for conducting physical, biological or scientific research.
- The information provided by remote sensing data may not be complete and may be temporary.
- Sometimes large scale engineering maps cannot be prepared from satellite data which makes remote sensing data collection incomplete.

Remote Sensing Platforms and Sensors

Platforms

Platform is a stage where sensor or camera is mounted to acquire information about a target under investigation. a platform is a vehicle, from which a sensor can be operated. Platforms can vary from stepladders to satellites. There are different types of platforms and based on its altitude above earth surface, these may be classified as:

Ground based Platforms

Wide varieties of ground-based platforms are used in remote sensing. Some of the common ones are hand held devices, tripods, towers and cranes. To study properties of a single plant or a small patch of grass, ground based platform is used. Ground based platforms (hand-held or mounted on a tripod) are also used for sensor calibration, quality control and for the development of new sensors.

For the field investigations, some of the most popular platforms have been used are 'cherry picker platform, portable masts and towers. The cherry picker platforms can be extended to approx. 15m. They have been used by various laboratories to carry spectral reflectance meters and photographic systems. Portable masts are also available in various forms and can be used to support cameras and sensors for testing. The main problem with these masts is that of stabilizing the platform, particularly in windy conditions. Permanent ground platforms like towers and cranes are used for monitoring atmospheric phenomenon and long-term monitoring of terrestrial features. Towers can be built on site and can be tall enough to project through a forest canopy so that a range of measurements can be taken from the forest floor, through the canopy and from above the canopy.

Balloon Platforms

Balloons as platforms are not very expensive like aircrafts. They have a great variety of shapes, sizes and performance capabilities. The balloons have low acceleration, require no power and exhibit low vibrations. There are three main types of balloon systems, viz. free balloons, Tethered balloons and Powered Balloons. Free balloons can reach almost top of the atmosphere; hence, they can provide a platform at intermediate altitude between those of aircraft and spacecraft.

Free floating or anchored balloons have altitude range of 22-40 km and can be used to a limited extent as a platform. It is used for probing the atmosphere and also useful to test the instrument under development.

Balloon as platform.

Aircraft Platform

Aerial platforms are primarily stable wing aircraft. Helicopters are also occasionally used for this purpose. Generally, aircraft are used to collect very detailed images. Helicopters can be for pinpoint locations but it vibrates and lacks stability.

Low Altitude Aircraft

It is most widely used and generally operates below 30,000 ft. They have single engine or light twin engine. It is suitable for obtaining image data for small areas having large scale.

Low altitude aircraft produced image.

Low Altitude Aircraft

It is more stable and operates above 30,000 ft. High altitude aircraft includes jet aircraft with good rate of climb, maximum speed, and high operating ceiling. It acquires imagery for large areas (smaller scale). Examples are NHAP, NAPP, AVIRIS. Aircraft platform acquire imagery under suitable weather conditions. It controls platform variables such as altitude. Time of coverage can also be controlled. However, it is expensive, less stable than spacecraft and has motion blurring.

Rockets as Platforms

Prior to use of airplanes, aerial photographs were obtained by rocketing a camera into the sky and then retrieving the camera and film. Synoptic imagery can be obtained from rockets for areas of some 500,000 square km. The Skylark earth Resource

Rocket is fired from a mobile launcher to altitudes between 90 - 400 kms. With the help of a parachute, the payload and the spent motor are returned to the ground gently thereby enabling speedy recovery of the photographic records. This rocket system has been used in surveys over Australia and Argentina. In 1946, V-2 rockets acquired from Germany after World War II were launched to high altitudes from White Sands, New Mexico. These rockets contained automated still or movie cameras that took picture as the vehicle ascended. The main problem with rockets is that they are one-time observations only. Except for one time qualitative or reconnaissance purposes, rocket platforms are not of much use in regular operational systems.

Spacecraft as Platform

Remote sensing is also conducted from the space shuttle or artificial satellites. Artificial satellites are manmade objects, which revolve around another object. The 1960s saw the primary platform used to carry remotely sensed instruments shifted from airplanes to satellite. Satellite can cover much more land space than planes and can monitor areas on a regular basis.

Beginning with the first television and infrared observation Satellite (tiRoS1) in 1960, early weather satellites returned rather poor views of cloud patterns and almost indistinct images of the earth's surface. Space photography becomes better and was further extended with the Apollo program. Then in 1973, Skylab the first American space workshop was launched and its astronauts took over 35,000 images of the earth with the earth Resources experiment Package (eReP) on board. Later on with LANADSAT and SPOT Satellite program, space photography received a higher impetus.

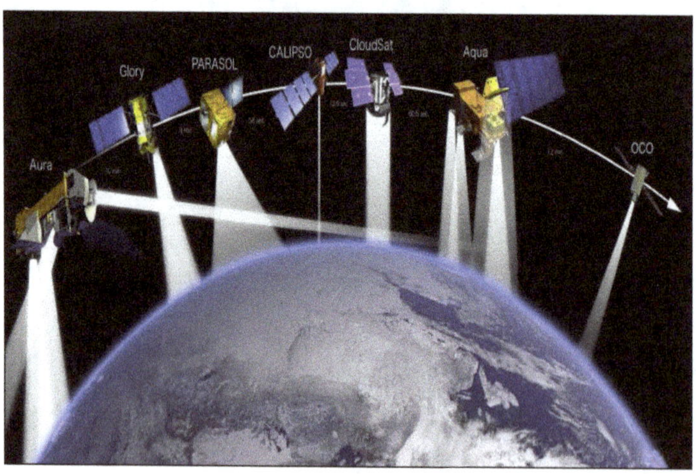

Types of Satellites

Here we will discuss about different types of satellites. You are familiar with the fact that today man-made or artificial satellites are widely used for a large number of purposes including military and civilian Earth observations, communication, navigation, weather forecasting and research purposes. Hence such satellites are classified into six major types namely, astronomical, communication, weather, remote sensing, navigation and reconnaissance satellites based on their uses.

Astronomical Satellites

An astronomical satellite is basically a big telescope floating in space. The satellite's vision is not clouded by the gases that make up the Earth's atmosphere because it is in orbit above the Earth. Therefore, astronomy satellites can see into the space up to ten times better than a telescope of similar strength on the Earth. Applications of astronomical satellites are mentioned below:

- To make star maps and take pictures of the planets in the solar system.
- To study mysterious phenomena such as black holes and quasars.
- Also to make maps of different planetary surfaces.

Astronomy satellites are different from space exploration satellites because they collect their data from Earth's orbit. In real sense, space exploration satellites are probes that are sent out into deep space. The most famous Hubble Space Telescope is an example of an astronomy satellite.

Communication Satellites

A satellite used for sending messages for telephonic conversations and conferencing, television broadcast, FAX, computer related network services or other communication/transmission to long distances is called a communication satellite. Communication satellites are commonly placed in the geostationary orbit. Now-a-days these satellites possibly form the greatest number of satellites that are in orbit. They are extensively used for the communication over large distances. It is really very difficult to go throughout a day without using a communication satellite at least once. You use communication satellites directly or indirectly while watching TV, in making long distance phone call, using a fax machine or even in listening to your favourite radio station. Communication satellites allow radio, television, and telephone transmissions to be sent live anywhere in the world. A communication satellite is a radio relay station in orbit above the Earth that receives, amplifies, and redirects analog and digital signals carried on a specific radio frequency.

Weather Satellites

As the name implies these satellites are used for the monitoring and forecasting of the weather. They can provide a much better understanding of weather phenomena. And, because of this technology, you can find out weather conditions anywhere in the world at any time of the day. These satellites provide meteorologists with scientific data to predict weather conditions and are equipped with advanced instruments. Weather satellites can give information about the followings:

- Radiation measurements from the Earth's surface and atmosphere which give information on amount of heat and energy being released from the Earth and its atmosphere.
- Satellites monitor the amount of snow in winter, the movement of ice fields in the Arctic and Antarctic regions, and the depth of the ocean.
- Infrared sensors on satellites examine crop conditions, areas of deforestation and regions of drought.

Earth Observation Satellites

Remote sensing is observing and measuring our Earth's environment from a distance. So, remote sensing or Earth observation satellites are usually put into space to monitor the resources that are important for humans. Remote sensing satellites are generally placed in the sun synchronous orbit. Using these satellites it is possible to see many features that are not obvious from the Earth's surface. For example, remote sensing satellites might track animal migration, locate mineral deposits and exploitation, watch agricultural crops for weather damage, etc.

All of these things can be done best from space because a satellite in orbit can Sensors and Platforms normally take photographs of large areas of land all over the world. Since these satellites are able to take images and observe areas all over the globe, the satellite is able to monitor areas in which the climate is very harsh, or which are nearly impossible to reach by land.

Navigation Satellites

Satellites for navigation were developed in the late 1950s as a direct result of ships needing to know exactly where they were at any given time. In the middle of the ocean or out of sight of land, you cannot find out your position accurately just by looking out of the window. In recent years, satellites have been used for accurate navigation. Global Positioning System (GPS) is the first system that was primarily intended for use as a highly accurate military system. Using GPS technology these satellites can provide exact location of a person on the Earth's surface. Since then it has been adopted by a huge number of commercial and private users. Now-a-days, GPS systems are available at reasonable costs that are affordable by individuals. GPS systems used for car navigation are good example of this. They are even being incorporated into phones in a system known as A-GPS (Assisted GPS) to enable better locational accuracy.

Reconnaissance Satellites

Reconnaissance satellites are used to spy on other countries. They provide intelligence information on the military activities of foreign countries. These satellites can even detect missile launches or nuclear explosions in space. Reconnaissance satellites can pick up and record radio and radar transmissions while passing over a country. There are following four types of reconnaissance satellites:

- Optical-imaging satellites that have light sensors that detect missile launches and locate enemy weapons on the ground.
- Radar-imaging satellites which are able to observe the Earth using radar technology through cloud cover.
- signals-intelligence or ferret satellites that are essentially super sophisticated radio receivers that capture the radio and microwave transmissions emitted from any country on the Earth.
- Relay satellites that make military satellite communications around the globe much faster by transmitting data from spy satellites to stations on the Earth.

Types of Orbits

Sun-Synchronous Orbits

In sun-synchronous orbit, satellite is orbiting in synchronous with sun. In this type of orbit, satellite passes every location at the same time each day. Such an orbit is near polar with altitudes between 300 and 1000 km. They are also known as "polar orbiters" because of the high latitudes they cross. The noon satellites pass over near noon and midnight, whereas morning satellites pass over near dawn and dusk.

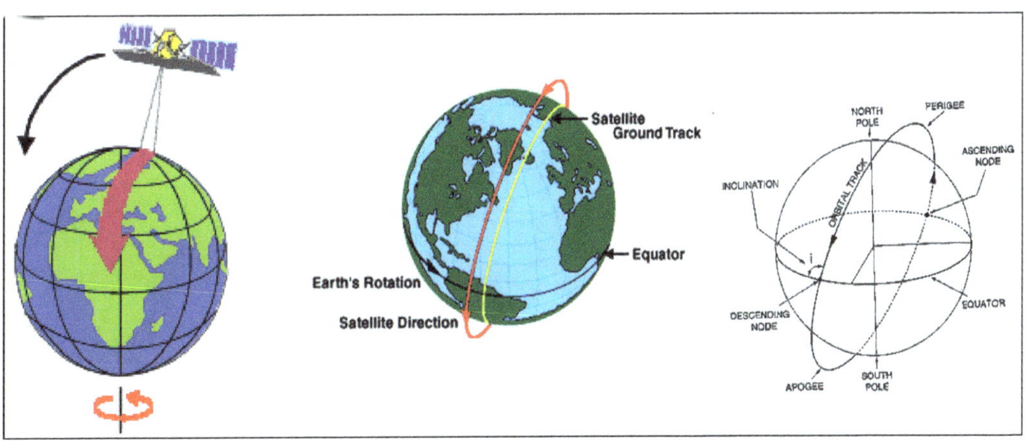

Satellite in a sun-synchronous orbit has an inclination that carries the satellite track westward at a rate that compensates for the change in local sun time as the satellite moves from north to south. Infact most earth observation satellites are placed in orbits designed to acquire imagery between 9.30 and 10.30 A.M. This local time is set to minimize cloud cover in tropical regions and provides optimum illumination. Satellites in sun-synchronous orbits pass from north to south on the sunlit side (descending node)

and from south to north on the shadowed side (the ascending node). During the descending pass, sensors that depend on reflected solar radiation acquire data, but radar and thermal sensors can acquire data independently of solar illumination, thereby observing the earth's surface during both passes.

The major advantage of such an orbit is that data taking is standardized with respect to the sun angle. Secondly, since the sun synchronous orbit is near polar, it provides nearly global coverage. Through these satellites, the entire globe is covered on regular basis and gives repetitive coverage on periodic basis. Some of them sun-synchronous satellites are LANDSAT Series, SPOT Series, IRS Series, NOAA, SEASAT, TIROS.

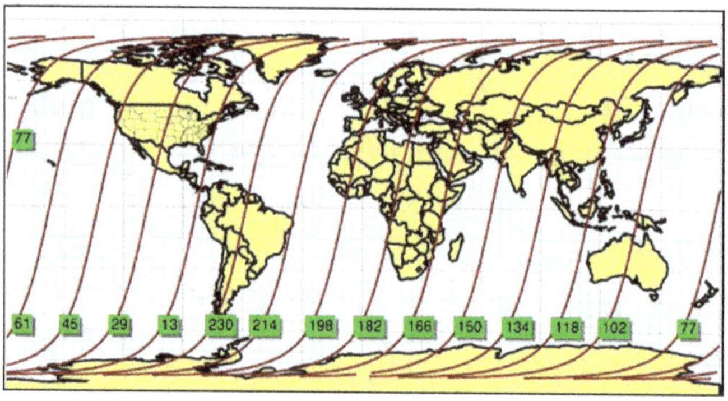

Geostationary (Geo-Synchronous) Orbits

In geosynchronous orbit, satellite in synchronous with earth. It is basically designed to maintain a constant position with respect to a specific portion of the earth's surface area day and night. These satellites move along in the same direction of the rotation of the earth, taking the same amount of time as the earth's rotation. Its coverage is limited to 70°N to 70°S latitudes and one satellite can view one-third globe. While moving west to east such satellites must have a velocity equal to that of the earth rotating about its axis west to east.

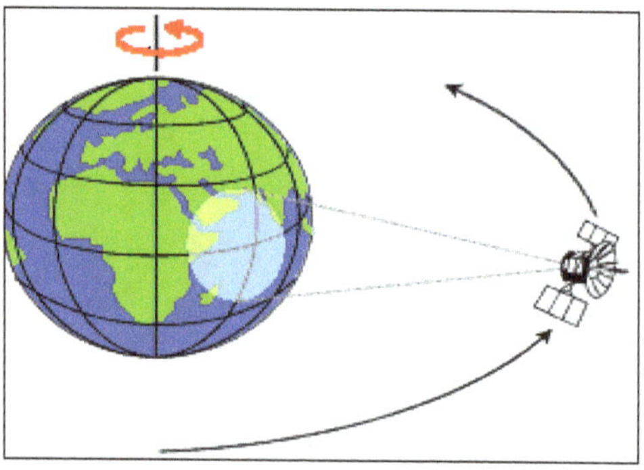

The main advantage of this satellite is that since it is stationary (rather appears stationary); it allows continuous viewing of that portion of the earth within the line of sight of the satellite sensors. However, satellite can only "see" one hemisphere. These are orbiting approximately 36,000 km above the Earth. Geostationary orbits are ideal for meteorological or communications satellites. Some examples of geostationary satellites are GOES, METOSAT, INTELSAT, INSAT.

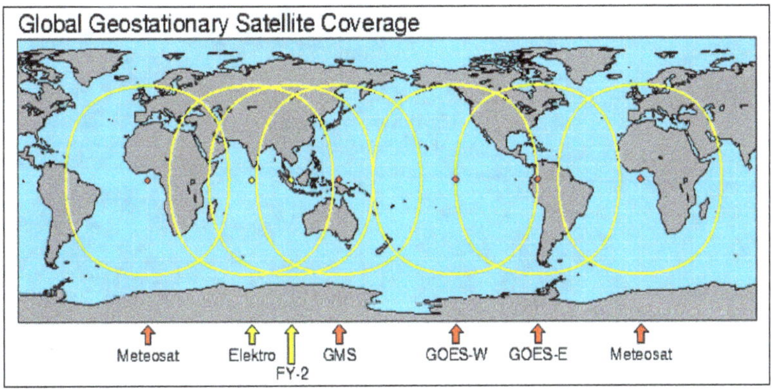

Sensors

Sensor is a device that gathers energy (EMR or other), converts it into a signal and presents it in a form suitable for obtaining information about the target under investigation. Remote sensors are mechanical devices, which collect information, usually in storable form, about objects or scenes, while being at some distance from them. Sensors used for remote sensing can be either those operating in Optical Infrared (OIR) region or those operating in the microwave region.

The first issue to be considered in selecting the sensor is the scale. Terrestrial and airborne sensors have high spatial resolution but sensors mounted on satellite, meant for acquiring synoptic coverage have relatively coarse spatial resolution. Selection of appropriate sensor system for the remote sensing satellite will also depend on the criteria to be addressed for using highly localised and refined or more regional and coarse sensor systems.

Sensor systems are classified as imaging and non-imaging sensors based on the type of output they provide. Imaging sensors measure the emitted/reflected intensity of EMR and provide image of the ground as output (e.g. photographic camera) whereas non-imaging sensors measures the intensity of radiation but does not provide any image and the observations are in the form of some numerical data (e.g. Gravimeter). We will focus here on the imaging sensor systems. As shown in the Fig, sensor systems can be broadly classified as passive or active systems based on the source of EMR:

- Passive Sensors: They detect the reflected or emitted EMR from natural sources. The useful wave bands are mostly in the visible and infrared region for passive remote sensing detectors.

- Active Sensors: They detect the reflected or emitted radiation from the objects which are irradiated from artificially generated energy sources, such as RADAR and LIDAR. The active sensor detectors are used in the radar and microwave regions.

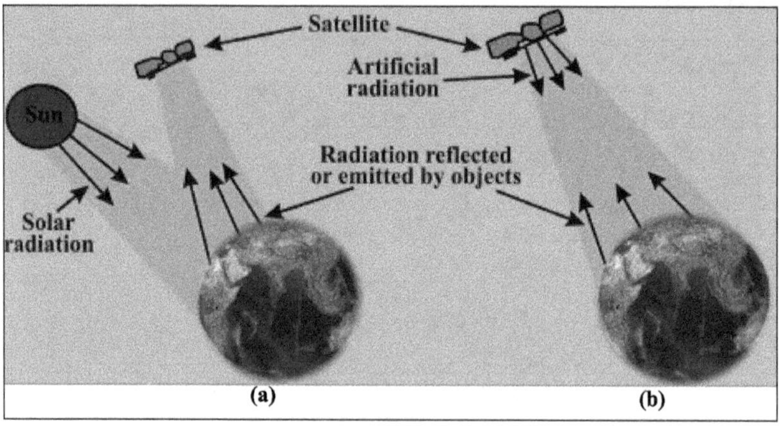

Schematics showing the functional mechanism of (a) passive and (b) active sensors.

Broadly, all the imaging sensors systems are classified based on technical components of the system and the capability of the detection by which the energy reflected by the terrain features is recorded.

The active and passive sensors can be further classified as shown in table.

Table: Types of remote sensing sensors.

Types of sensors	Subtypes based on the wavelength region of EMS detected	Subtypes based on sensing mechanism	Examples
Passive sensors	Optical-infrared (multispectral) sensors	Photographic and television cameras	MKF-6M; S065; Bhaskara-1, 2; Landsat RBV.
		Opto-mechanical scanner	Landsat MSS and TM, INSAT, VHRR
		Push-broom scanner	IRS-LISS, WiFS SPOT HRV
	Thermal sensors	Thermal infrared scanner	Landsat TM, TIMS, MODIS, ASTER, AVHRR
	Microwave sensors	Scanning microwave radiometer	Oceansat-1 MSMR (Multi-channel scanning microwave)
Active sensors	Optical-infrared sensors	LIDAR	Leica ALS70 Airborne LIDAR Sensor
	Microwave sensors	RADAR	RADARSAT SAR
		Scatterometer/Altimeter	Seasat

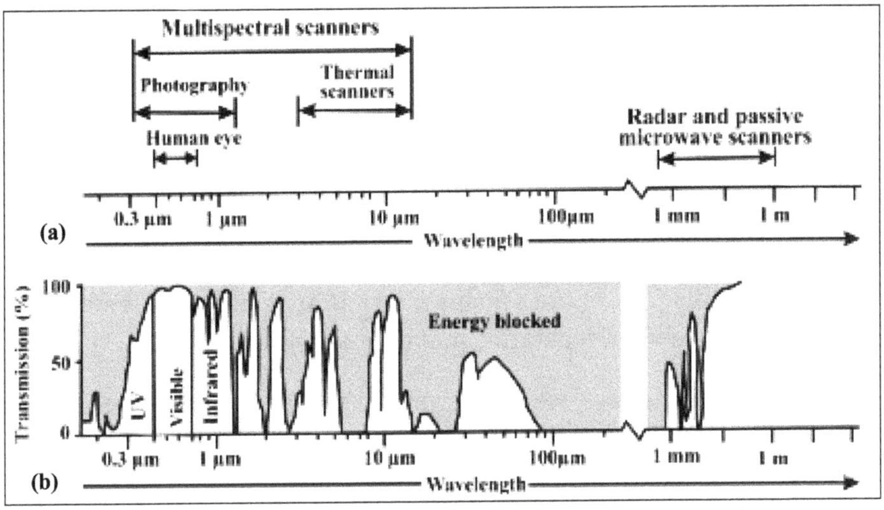

Atmospheric transmission and the wavebands of common remote sensing systems: (a) Remote sensing scanners, and (b) atmospheric transmittance.

Multispectral Imaging Sensor Systems

The multispectral imaging sensors include photographic and scanning systems. The photographic system is an imaging system in which cameras are used. In the scanning system, scanners along with filters for various wavelength regions are used. In some cases both photographic and scanning systems are used in combination.

In the photographic system, different parts of the spectrum are sensed with different film-filter combinations and images are formed directly on to a film. In the opto-mechanical (scanning system) sensors, the optical image is first converted into an electrical signal (video data) and later processed to record or transmit the data. But the photographic system suffers from one major defect of considerable distortion at the edges. This is due to large lens opening.

Analog (Photographic) Systems

Photographic cameras are the oldest and most widely used remote sensing sensor especially in aerial photography. Cameras belong to framing system; have been successfully used as the remote sensors from aircraft, balloons and manned and unmanned spacecraft. Camera systems are passive optical sensors that use a lens or system of lenses to form an image at the focal plane where an image is sharply defined. The images are recorded on the photographic films which are sensitive to light from 0.3 ìm to 0.9 ìm in wavelength covering the ultraviolet, visible, and near-infrared. The analog or photographic film is generally composed of single layer of silver halide emulsion (black and white or panchromatic) or triples layers of yellow, magenta and cyan colours for colour photographs. When the photographic film is exposed, over a particular area the molecules of the film's emulsions react with the radiation emitting from that particular area. And the film developed produces a black and white or colour representation of the continuous range of light intensity as evidenced across that particular area.

Box-Framing systems instantaneously measure radiation coming from the entire scene at once and record an image of an area. Cameras, eye and vidicons are common examples of framing systems. The size of the scene that is framed depends on the apertures and optics of the system that define the field of view. Generally, the panchromatic and colour films are sensitive to the light in the visible region. But the photographic films developed specially for remote sensing analysis are sensitive to reflected-IR light, which are used extensively for vegetative mapping analysis.

Earlier, photographic imageries were interpreted visually. Now-a-days the scanning technology made it easier to convert whole analog photographs to digital images. The scanned photographic imageries take advantage of feature enhancement and extraction facilities available within most image processing software packages. Nevertheless, the spatial and spectral resolution of the original photograph cannot be enhanced by scanning.

Scanning Systems

Photographic system uses chemical reactions on the surface of a light sensitive film to detect energy variations within a scene and the film acts as both detecting and recording medium whereas the opto-mechanical sensor generates an electrical signal that corresponds to the energy variations in the original scene and the signals are recorded onto some magnetic medium.

The scanning systems use an electrical sensor called detector that records the brightness of the small scene of the terrain within its instantaneous field-of-view (IFOV) to produce an image. The brightness (electrical) signals which are recorded by the detector vary continuously in proportion as the optics of the sensors and mirrors sweep the IFOV over the landscape. This signal is amplified, recorded in the magnetic tape and then converted to digital form to produce an image. In this way an image is produced from a series of adjacent cells or picture elements (pixels). The scanning system is able to record the brightness of the entire terrain by sweeping the detector rapidly across the terrain.

Multispectral scanner systems can sense from 0.3 μm to 14 μm and further they can sense in very narrow bands. According to lens theory, the distortions can be minimised and resolution be improved considerably by using a narrow beam of light and this can be achieved by a system called scanning system. The multispectral scanning systems sweep the IFOV of the detector across the terrain in a series of parallel scan lines. There are four common types of scanning modes for the scanning system:

- Across-track scanning system
- Along-track scanning system

- Side looking or oblique scanning system (radar)
- Spin scanning system

Across-Track Scanners

This type of scanning system employs a faceted mirror that is rotated by an electric motor in which horizontal axis of rotation aligned parallel with the direction of flight. The mirror scans the terrain in a pattern of parallel scan lines that are right angles to the direction of the airborne platform as shown in Fig. Energy reflected or radiated from the ground is reached onto the detector by the mirrors. This kind of scanner systems is also called as whisk broom scanner system. The factors affecting the strength of sensor signal produced by a detector are the energy flux, altitude of the sensor, spectral bandwidth of the detector, IFOV and dwell time. The across-track scanners have short dwell time that results in detector receiving less energy and creating a weaker signal. Examples of across-track scanners are Multispectral Scanner (MSS) and Thematic Mapper (TM) of Landsat series of satellites.

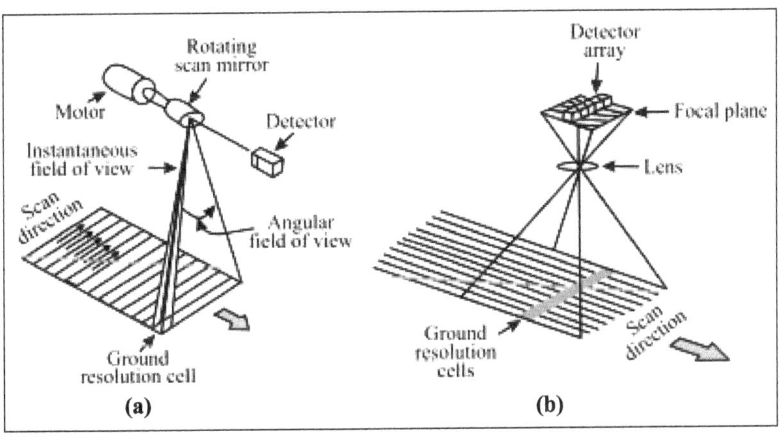

Schematics showing sensor's scanning modes: (a) across-track mode; and (b) along-track mode.

Landsat Multispectral Scanner (MSS)

It was the primary sensor system for the Landsats 1-3 and 4-5. This sensor had four spectral bands ranging from 0.5 to 1.1 μm in the electromagnetic spectrum that record reflected radiation from the Earth's surface. These bands are green (0.5 to 0.6 μm), red (0.6 to 0.7 μm), and near-infrared (0.7 to 0.8 and 0.8 to 1.1 μm). In addition, the subsequent Landsat MSS contained an additional thermal infrared band of electromagnetic spectrum. MSS had an oscillating mirror that scans a 185 km wide ground swath. MSS also had a spectrometer and six detectors for each spectral band. The spectrometer separates the solar radiation reflected from the landscape into four different spectral bands or wavelengths whereas detectors record six scan lines during every sweep of the scan mirror. Later, the energy sensed by the detectors is converted into electrical signals for recording and transmission as an image data.

Landsat data are widely used for detecting and monitoring Earth's resources. Among the four bands of MSS, band 1 is used to detect green reflectance from healthy vegetation, and band 2 for detecting chlorophyll absorption in vegetation. MSS bands 3 and 4 are ideal for recording near infrared reflectance peaks in healthy green vegetation and for detecting water-land interfaces.

Thematic Mapper

Thematic Mapper (TM) is an advanced second generation of MSS first deployed on Landsat-4 and 5. Its design offered spatial, radiometric, and geometric improvements over the MSS systems. The design of the TM was more complicated than the MSS. TM is an across-track scanner similar to MSS with an oscillating scan mirror and arrays of detectors. TM provides data which are scanned simultaneously in seven narrow spectral bands covering visible (blue- 0.45 - 0.52 ìm, green - 0.52 - 0.60 ìm, and red - 0.63 - 0.69 ìm), near infrared (0.76 - 0.90 ìm), middle infrared (1.55 - 1.75 ìm and 2.08 - 2.35 ìm) and thermal infrared (10.4 - 12.5 ìm). TM provides data with a 30 m resolution in the visible, near infrared and middle infrared and 120 m resolution in the thermal infrared regions of EMR.

Along-Track Scanning System

Along-track scanners record multiband image data along a swath beneath the aircraft. As the aircraft advances in the forward direction, the scanner scans the Earth with respect to the designed swath to build a two dimensional images by recording successive scan lines that are oriented at right angles to the direction of aircraft. In this system, detectors are placed in a linear array in the focal plane of the image formed by a lens system. This system is also called as a push broom system because the array of the detectors that record terrain brightness along a line of pixels are effectively pushed like a broom along the orbital path of the aircraft. The along track scanners have long dwell time in which the detectors provide fine spatial and higher spectral resolutions.

The difference between along-track system and across track system is that the linear arrays of detectors are used in push broom sensor instead of a rotating mirror. Example of an along-track scanner is SPOT-High Resolution Visible (HRV) camera and Linear Imaging Self Scanning Sensors (LISS) of Indian remote sensing (IRS).

Linear Imaging Self-Scanning Sensor

Linear imaging self-scanning (LISS) is a sensor system designed by ISRO for Indian remote sensing satellites. In fact, LISS is a multispectral camera system in which each camera system contains four imaging lens assemblies one for each band followed by a linear charge coupled devices (CCD) array. The optics focuses a strip of landscape in the across track direction on to the sensor array. The images obtained from each detector are stored and later shifted out to get video signals. It operates in four bands which are

B1 (0.45-0.52 μm), B2 (0.52-0.59 μm), B3 (0.62-0.68 μm) and B4 (0.77-0.86 μm) in the visible and near infrared wavelength of electromagnetic region. At present, there are four versions of LISS and each version has its own specifications.

LISS-III is a medium resolution multispectral camera almost similar to the LISS-I. In LISS-III camera, the band 1 is replaced by short-wave infrared (SWIR) band B5 (1.55-1.70 μm). As a consequence, it provides images in visible as well as SWIR bands. The spatial resolution and ground swath of LISS-III is 23.5 m and 141 km, respectively, in all the four bands. LISS-IV is a high resolution multispectral camera operating in three spectral bands (B2, B3, B4). LISS-IV camera operates in both mono (any single band out of B1, B2 and B3) and multispectral mode. The ground swath of LISS-IV is 23.9 km and 70 km for multispectral and mono mode, respectively. LISS-IV camera can be tilted up to ± 26° in the across track direction thereby providing a revisit period of 5 days.

Side Looking or Oblique Scanning Systems (Radar)

Side looking scanning system is an active scanning system, e.g. radar. This system itself generates EMR, then illuminates the terrain and detects the energy (radar pulses) returning back from the terrain and records it as an image. Therefore, radar imagery is received by collecting and measuring the reflections of pulses sent out from radar equipped in the aircraft and satellites. Side Looking Airborne Radar (SLAR) is one of the common types of remote sensing techniques used for obtaining radar images of the terrain. The main components of SLAR include antenna, duplexer, transmitter, receiver, pulse-generating device and cathode ray tube.

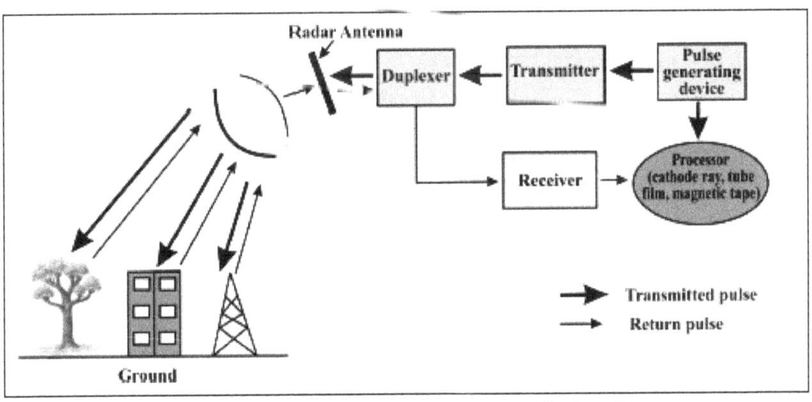

Schematics showing components of a SLAR.

In this system, the radar antenna sends the radar pulse to the ground and receives the radar return from the ground. Duplexer is an electronic switch and its main function includes the prevention of interference between radar return and transmitted beams. Receiver records the timing and intensity of radar return and also amplifies the weak pulses received by the antenna. This helps to identify features of the terrain appeared on an image. Finally, radar return may be displayed on a cathode ray tube and get recorded on film or in the magnetic tapes. Overall, an image created is a function of time and strength of the radar pulse that is returned from the Earth's objects. The resolution of radar systems can be determined by the radar beam width of the microwave pulse generated by the system. It is calculated by the equation given below:

Resolution (m) = range (km) × wavelength (cm) / antenna aperture (m)

The range direction is the direction of the pulses of the microwave transmitted from the antenna to illuminate the strips of the landscape in the look direction. While look direction is perpendicular to the azimuth or aircraft flight direction. Range direction may be near or far range. In near range, the transmitted pulses lie close to the aircraft flight path and hence have short travel time but in far range, pulses are very far away from the aircraft and have long travel time.

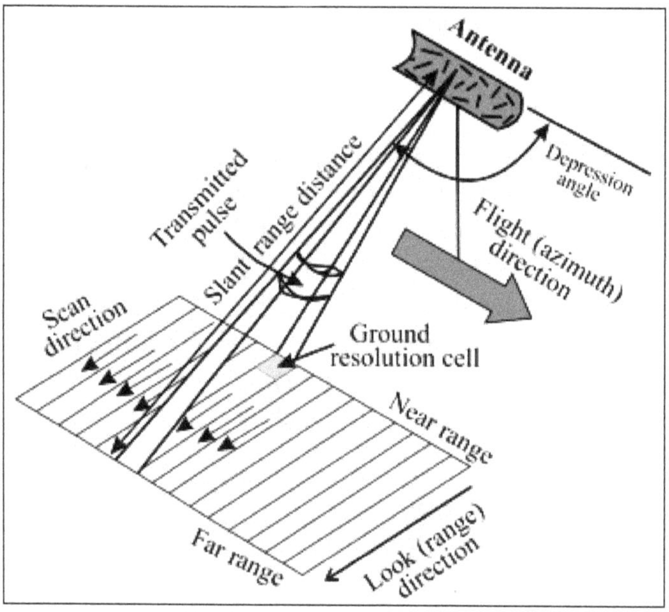

Side looking scanning system.

Thermal Remote Sensing System

Thermal scanners belong to the electro-optical scanning systems. These scanners sense the thermal infrared portion of the EMS. Thermal scanners do not record the true internal temperature of objects but record the pattern of radiant temperature variation of the objects. As a consequence, they sense energy emitted rather than reflected from objects, therefore, thermal scanners can operate day or night. Thermal scanners use photo detectors to detect emitted thermal radiation. The detectors are cooled to temperatures close to absolute zero in order to limit their own thermal emissions.

Thermal infrared scanners scan the terrain in the across-track mode. A thermal airborne infrared scanner consists of an electric motor and a rotating shaft which is oriented parallel to the aircraft flight direction. Both these instruments are mounted in the aircraft. The scan mirror which is inclined at 45^0 mounted on the one end of the shaft sweeps the terrain at a right angle to the fight path. This scan mirror also detects the infrared energy emitted from the terrain. Later, it sends the energy to focusing mirrors where the energy get detected by the detector. The detector converts the emitted energy into an electrical signal. The signal varies in proportion according to the intensity of emitted infrared radiation. Detector is normally placed by a vacuum bottle filled with liquid nitrogen. A second mirror known as recorder mirror placed at another end of the shaft rotates synchronously with the scan mirror and sweeps the image of the modulated light source across a strip of recording film.

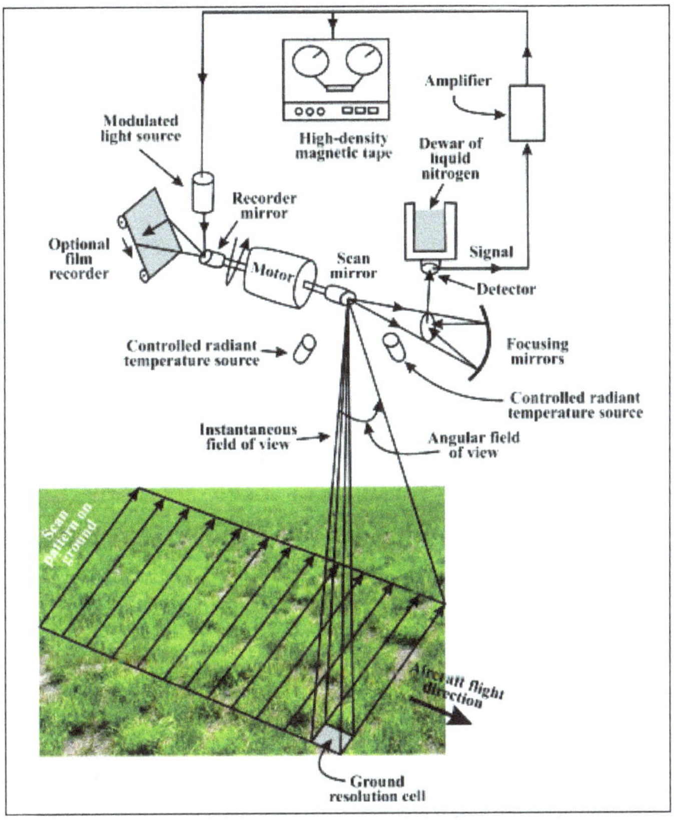

Schematic representation of thermal infrared scanner.

Heat Capacity Mapping Mission (HCMM) detected radiation in two bands which were visible/near- infrared (0.5-1.1 ì m) and thermal infrared (10.5-12.5 ì m). The Landsat-3 MSS and TM both included thermal bands of EM spectrum. Since Landsat TM has seven spectral bands hence provides more radiometric information than Landsat MSS. Thermal Infrared Multispectral Scanner (TIMS) which was developed by the Jet Propulsion Laboratory of NASA collects data with six thermal bands ranging from 8.2 to 12.2 ìm. Moderate Resolution Imaging Spectroradiometer (MODIS) was launched in 1999 with 36 spectral bands and spatial resolution ranging from 250-1,000 m. Advanced Spaceborne Thermal Emission and Reflection Radiometer (ASTER) has 14 spectral bands and scans the terrain in visible/NIR, SWIR and thermal infrared with spatial resolutions 15 m, 30 m, and 90 m, respectively. Advanced Very High Resolution Radiometer (AVHRR) of the National

Oceanic and Atmospheric Administration (NOAA) has thermal infrared scanners in the two thermal atmospheric windows (3.55-3.93 ìm and 10.5-11.5 ìm) with a ground resolution of 1.1 km to measure top cloud temperature and seawater temperatures.

Thermal sensors essentially measure the surface temperature and thermal properties of objects. Hence, thermal imagery has diverse application in the field of geology (geologic mapping of the formation with different thermal properties and for locating underground coal mine fires), soil science (soil mapping, and determination of moisture conditions), and forestry (studying evapo-transpiration from vegetation, delineating the extent of active forest fires) and for the study urban land surface temperature (detection of heat losses from houses, buildings and factories).

Microwave Imaging System

The microwave region of the EM spectrum includes wavelengths from 1mm to 1 m. The advantage of microwave remote sensing is that microwaves are capable of penetrating the atmosphere when it is with conditions like cloud cover, snow and smoke. They also have the capacity of sensing in day or night. Microwave imaging systems can be classified into two categories namely, a) active and b) passive microwave remote sensing.

- Active Microwave Remote Sensing: Active microwave sensing systems are of two types and they are imaging sensors and non-imaging sensors. Most imaging sensors or imaging radars used for remote sensing e.g. SLAR. These imaging radars are again divided into real aperture and synthetic aperture systems.

- Passive Microwave Remote Sensing: Non-imaging remote sensing radars are either scatterometers or altimeters. Passive microwave sensors, called radiometers, measures the natural emitted energy from the Earth's surface. The suitable antenna collects the emitted energy and transforms it as a signal. It is represented as an equivalent temperature, i.e. the temperature of a black body source which produces the same amount of signal in bandwidth of the system. Passive remote sensing is possible if the radiometer is used in a scanning mode just like the optical scanner.

Remote Sensing in Solar Reflection Region

Radiation from the sun, when incident upon the earth's surface, is reflected by the surface, transmitted into the surface or absorbed and emitted by the surface. The EMR, on interaction, experiences a number of changes in magnitude, direction, wavelength, polarization and phase. These changes are detected by the remote sensor and enable the interpreter to obtain useful information about the object of interest. The remotely sensed data contain both spatial information (size, shape and orientation) and spectral information (tone, colour and spectral signature).

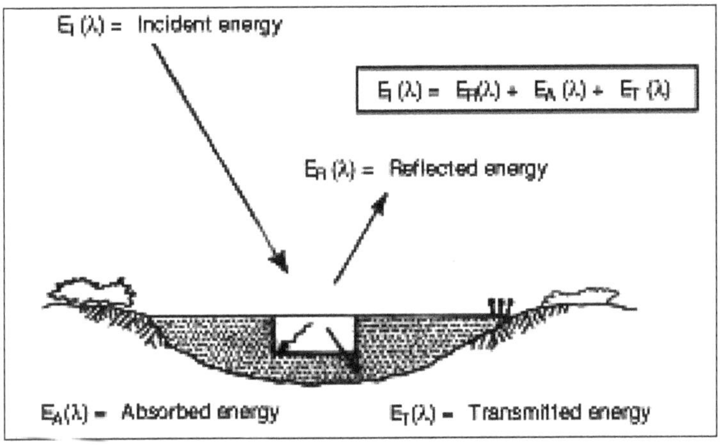

From the viewpoint of interaction mechanisms, with the object-visible and infrared wavelengths from 0.3 μm to 16 μm can be divided into three regions. The spectral band from 0.3 μm to 3 μm is known as the reflective region. In this band, the radiation sensed by the sensor is that due to the sun, reflected by the earth's surface. The band corresponding to the atmospheric window between 8 μm and 14 μm is known as the thermal infrared band. The energy available in this band for remote sensing is due to thermal emission from the earth's surface. Both reflection and self-emission are important in the intermediate band from 3 μm to 5.5 μm.

In the microwave region of the spectrum, the sensor is radar, which is an active sensor, as it provides its own source of EMR. The EMR produced by the radar is transmitted to the earth's surface and the EMR reflected (back scattered) from the surface is recorded and analyzed. The microwave region can also be monitored with passive sensors, called microwave radiometers, which record the radiation emitted by the terrain in the microwave region.

Reflection

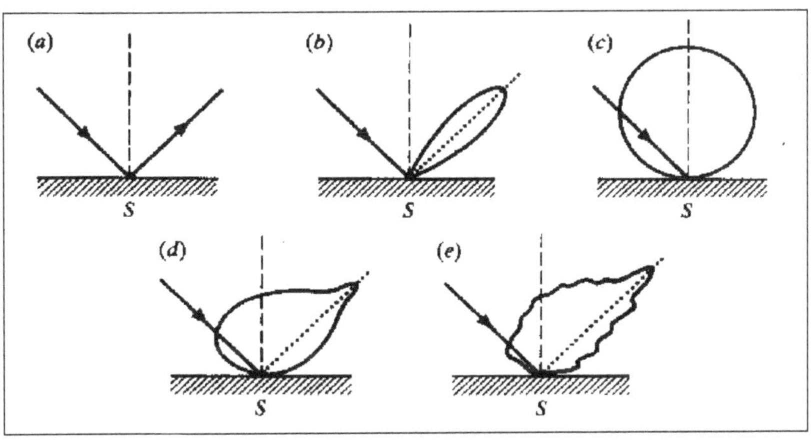

Different types of scattering surfaces: (a) Perfect specular reflector (b) Near perfect specular reflector (c) Lambertian (d) Quasi-Lambertian (e) Complex.

Of all the interactions in the reflective region, surface reflections are the most useful and revealing in remote sensing applications. Reflection occurs when a ray of light is redirected as it strikes a non-transparent surface. The reflection intensity depends on the surface refractive index, absorption coefficient and the angles of incidence and reflection.

Transmission

Transmission of radiation occurs when radiation passes through a substance without significant attenuation. For a given thickness, or depth of a substance, the ability of a medium to transmit energy is measured as transmittance (τ).

Spectral Signature

Spectral reflectance, [$\rho(\lambda)$], is the ratio of reflected energy to incident energy as a function of wavelength. Various materials of the earth's surface have different spectral reflectance characteristics. Spectral reflectance is responsible for the color or tone in a photographic image of an object. Trees appear green because they reflect more of the green wavelength. The values of the spectral reflectance of objects averaged over different, well-defined wavelength intervals comprise the spectral signature of the objects or features by which they can be distinguished. To obtain the necessary ground truth for the interpretation of multispectral imagery, the spectral characteristics of various natural objects have been extensively measured and recorded. The spectral reflectance is dependent on wavelength, it has different values at different wavelengths for a given terrain feature. The reflectance characteristics of the earth's surface features are expressed by spectral reflectance, which is given by:

$$\rho(\lambda) = \left[ER(\lambda) / EI(\lambda)\right] \times 100$$

Where,

$\rho(\lambda)$ = Spectral reflectance (reflectivity) at a particular wavelength

$ER(\lambda)$ = Energy of wavelength reflected from object

$EI(\lambda)$ = Energy of wavelength incident upon the object

The plot between $\rho(\lambda)$ and λ is called a spectral reflectance curve.

This varies with the variation in the chemical composition and physical conditions of the feature, which results in a range of values. The spectral response patterns are averaged to get a generalized form, which is called generalized spectral response pattern for the object concerned. Spectral signature is a term used for unique spectral response pattern, which is characteristic of a terrain feature. The spectral characteristics of the three main earth surface features are discussed below:

- Vegetation: The spectral characteristics of vegetation vary with wavelength.

Plant pigment in leaves called chlorophyll strongly absorbs radiation in the red and blue wavelengths but reflects green wavelength. The internal structure of healthy leaves acts as diffuse reflector of near infrared wavelengths. Measuring and monitoring the near infrared reflectance is one way that scientists determine how healthy particular vegetation may be.

- Water: Majority of the radiation incident upon water is not reflected but is either absorbed or transmitted. Longer visible wavelengths and near infrared radiation is absorbed more by water than by the visible wavelengths. Thus water looks blue or blue green due to stronger reflectance at these shorter wavelengths and darker if viewed at red or near infrared wavelengths. The factors that affect the variability in reflectance of a water body are depth of water, materials within water and surface roughness of water.
- Soil: The majority of radiation incident on a soil surface is either reflected or absorbed and little is transmitted. The characteristics of soil that determine its reflectance properties are its moisture content, organic matter content, texture, structure and iron oxide content. The soil curve shows less peak and valley variations. The presence of moisture in soil decreases its reflectance.

By measuring the energy that is reflected by targets on earth's surface over a variety of different wavelengths, we can build up a spectral signature for that object. And by comparing the response pattern of different features we may be able to distinguish between them, which we may not be able to do if we only compare them at one wavelength. For example, Water and Vegetation reflect somewhat similarly in the visible wavelength but not in the infrared.

Interactions with the Atmosphere

The sun is the source of radiation, and electromagnetic radiation (EMR) from the sun that is reflected by the earth and detected by the satellite or aircraft-borne sensor must pass through the atmosphere twice, once on its journey from the sun to the earth and second after being reflected by the surface of the earth back to the sensor. Interactions of the direct solar radiation and reflected radiation from the target with the atmospheric constituents interfere with the process of remote sensing and are called as "Atmospheric Effects".

The interaction of EMR with the atmosphere is important to remote sensing for two main reasons. First, information carried by EMR reflected/emitted by the earth's surface is modified while traversing through the atmosphere. Second, the interaction of EMR with the atmosphere can be used to obtain useful information about the atmosphere itself.

The atmospheric constituents scatter and absorb the radiation modulating the radiation reflected from the target by attenuating it, changing its spatial distribution and introducing into field of view radiation from sunlight scattered in the atmosphere and some of the energy reflected from nearby ground area. Both scattering and absorption vary in their effect from one part of the spectrum to the other. The solar energy is subjected to modification by several physical processes as it passes the atmosphere, viz. scattering; Absorption, and Refraction.

Atmospheric Scattering

Scattering is the redirection of EMR by particles suspended in the atmosphere or by large molecules of atmospheric gases. Scattering not only reduces the image contrast but also changes the spectral signature of ground objects as seen by the sensor. The amount of scattering depends upon the size of the particles, their abundance, the wavelength of radiation, depth of the atmosphere through which the energy is traveling and the concentration of the particles. The concentration of particulate matter varies both in time and over season. Thus the effects of scattering will be uneven spatially and will vary from time to time.

Atmospheric Absorption

The gas molecules present in the atmosphere strongly absorb the EMR passing through the atmosphere in certain spectral bands. Mainly three gases are responsible for most of absorption of solar radiation, viz. ozone, carbon dioxide and water vapour. Ozone absorbs the high energy, short wavelength portions of the ultraviolet spectrum ($\lambda < 0.24$ μm) thereby preventing the transmission of this radiation to the lower atmosphere. Carbon dioxide is important in remote sensing as it effectively absorbs the radiation in mid and far infrared regions of the spectrum. It strongly absorbs in the region from about 13-17.5 μm, whereas two most important regions of water vapour absorption are in bands 5.5 - 7.0 μm and above 27 μm. Absorption relatively reduces the amount of light that reaches our eye making the scene look relatively duller.

Refraction

The phenomenon of refraction, that is the bending of light at the contact between two media, also occurs in the atmosphere as the light passes through the atmospheric layers

of varied clarity, humidity and temperature. These variations influence the density of atmospheric layers, which in turn, causes the bending of light rays as they pass from one layer to another. The most common phenomena are the mirage like apparitions sometimes visible in the distance on hot summer days.

3

Cartography and Map Making

The art and practice of representing a geographical area on a map or chart is known as cartography. In order to completely understand cartography, it is necessary to understand the processes related to it. The following chapter elucidates the varied processes and mechanisms associated with this area of study like map projection, map designing and map algebra.

Cartography

Cartography is a complex, an ever-changing field, but at the center of it is the map-making process. Viewed in the broadest sense, this process includes everything from the gathering, evaluation and processing of source data, through the intellectual and graphical design of the map, to the drawing and reproduction of the final document. As such, it is a unique mixture of science, art and technology and calls for a variety of in-depth knowledge and skills on the part of the cartographer. Sometimes one person directs this entire sequence of cartographic activities, but this occurs only in relatively simple cases. In the creation of a map, it is much more common for the various tasks to be split up and accomplished by several individuals.

Cartography is much more than just map-making, however. It is also an academic discipline in its own right. It has its own professional associations (regional, national and international), journals, conferences, educational programs and its own identity. As a discipline, it embraces not only cartographers, who make maps, but also cartographers who teach about maps and cartographers who do research on maps. Once seen as the products of a relatively straightforward practical exercise, maps are now viewed as complex intellectual images offering a rich potential for scientific investigation. Whether the thrust of the research is cognitive, mathematical, historical, perceptual or technological, cartographers are exploiting this potential to the fullest.

Cartography today has two essential characteristics. First of all, it is important. Maps perform a fundamental and indispensable role as one of the underpinnings of civilization. Few activities relating to the earth's surface, whether land use planning, property ownership, weather forecasting, road construction, locational analysis, emergency response, forest management, mineral prospecting, navigation–the list is endless–would be practicable without maps. And never has this role been more vital than it is today. Humanity faces severe problems, many of them environmental in nature, and effective mapping is crucial if solutions are to be found. In conjunction with the great data gathering capabilities and analytical power of remote sensing and geographic information systems (GIS), cartography, in many instances, provides the key to finding solutions.

A second, overriding characteristic of cartography today is its dynamic nature. The cartographic discipline is in the throes of a revolution, brought about in large measure by advances in technology and, in particular, by the impact of the computer. Map-making has always been affected by technological change, but the recent transformation of cartography has been unprecedented in scope compared with previous evolutionary changes. While former pen and ink techniques for map-making are still found in isolated use, today's cartographer is more likely to be found seated at a terminal using the very latest in computer hardware and graphic software. In most cases, without any loss in image quality, maps are generated faster with less cost than before, not to mention with even more enjoyment for the map-maker. In essence, the computer equips the cartographer with unparalleled control over the mapping process.

Essence of Cartography

Atlas

An atlas is a collection of various maps of the earth or a specific region of the earth, such as the U.S. or Europe. The maps in atlases show geographic features, the topography of an area's landscape and political boundaries. They also show climatic, social, religious and economic statistics of an area. Maps that make up atlases are traditionally bound as books. These are either hardcover for reference atlases or softcover for atlases that are meant to serve as travel guides. There are also countless multimedia options for atlases, and many publishers are making their maps available for personal computers and the Internet.

The use of maps and cartography to understand the world has a very long history. It is believed that the name "atlas," meaning a collection of maps, came from the mythological figure Atlas. Legend says that Atlas was forced to hold the earth and the heavens on his shoulders as a punishment from the gods. His image was often printed on books with maps and they eventually became known as atlases. The earliest known atlas is associated with the Greco-Roman geographer Claudius Ptolemy. His work, Geographia, was the first published book of cartography, consisting of the knowledge of the world's geography that was known around the time of the second century. Maps and manuscripts were written by hand at the time. Geographia's earliest surviving publications date back to 1475.

The voyages of Christopher Columbus, John Cabot, and Amerigo Vespucci increased knowledge of the world's geography in the late 1400s. Johannes Ruysch, a European cartographer and explorer, created a new map of the world in 1507 that became very popular. It was reprinted in a Roman edition of Geographia that year. Another edition of Geographia was published in 1513 and it connected North and South America.

The first modern atlas was printed in 1570 by Abraham Ortelius, a Flemish cartographer and geographer. It was called Theatrum Orbis Terrarum, or Theater of the World. It was the first book of maps with images that were uniform in size and design. The first edition consisted of 70 different maps. Like Geographia, Theater of the World was extremely popular and it was printed in numerous editions from 1570 to 1724. In 1633, a Dutch cartographer and publisher named Henricus Hondius designed an ornately decorated world map that appeared in an edition of Flemish geographer Gerard Mercator's atlas, originally published in 1595.

The works by Ortelius and Mercator are said to represent the beginning of the Golden Age of Dutch cartography. This is the period when atlases grew in popularity and became more modern. The Dutch continued to produce many volumes of atlases throughout the 18th century, while cartographers in other parts of Europe also began to print their works. The French and British began to produce more maps in the late 18th century, as well as sea atlases because of their increased maritime and trade activities.

By the 19th century, atlases began to get very detailed. They looked at specific areas such as cities instead of whole countries and/or regions of the world. With the advent of modern printing techniques, the number of atlases published also began to increase. Technological advances such as Geographic Information Systems (GIS) have allowed modern atlases to include thematic maps that show various statistics of an area.

Types of Atlases

Because of the wide variety of data and technologies available today, there are many different types of atlases. The most common are desk or reference atlases, and travel atlases or roadmaps. Desk atlases are hardcover or paperback, but they're made like reference books and they include a variety of information about the areas they cover. Reference atlases are generally large and include maps, tables, graphs and other images and text to describe an area. They can be made to show the world, specific countries, states or even specific locations such as a national park. The National Geographic Atlas of the World includes information about the entire globe, broken down into sections that discuss the human world and the natural world. These sections include the topics of geology, plate tectonics, biogeography, and political and economic geography. The atlas then breaks the world down into continents, oceans and major cities to show political and physical maps of the continents as a whole and the countries within them. This is a very large and detailed atlas, but it serves as a perfect reference for the world with its many detailed maps as well as images, tables, graphs, and text.

The Atlas of Yellowstone is similar to the National Geographic Atlas of the World but it's less extensive. This, too, is a reference atlas, but instead of examining the entire world, it looks at a very specific area. Like the larger world atlas, it includes information on the human, physical and biogeography of the Yellowstone region. It offers a variety of maps that show areas within and outside of Yellowstone National Park.

Travel atlases and roadmaps are usually paperback and are sometimes spiral bound to make them easier to handle while traveling. They often do not include all the information that a reference atlas would, but instead focus on information that may be useful to travellers, such as specific road or highway networks, the locations of parks or other tourist spots, and, in some cases, the locations of specific stores and/or hotels. The many different types of multimedia atlases available can be used for reference and/or travel. They contain the same types of information you'd find in book format.

Globe

Globe, sphere or ball that bears a map of the Earth on its surface and is mounted on an axle that permits rotation. The ancient Greeks, who knew the Earth to be a sphere, were the first to use globes to represent the surface of the Earth. Crates of Mallus is said to have made one in about 150 BCE. In ancient times, globes also were used to represent the constellations; the earliest surviving globe is the marble Farnese globe, a celestial globe dating from about 25 CE.

Today's globe, typically hollow, may be made of almost any light, strong material, such as cardboard, plastic, or metal. Some are translucent. They may also be inflatable. Terrestrial globes are usually mounted with the axis tilted 23.5° from the vertical, to help simulate the inclination of the Earth relative to the plane in which it orbits the Sun. Terrestrial globes may be physical, showing natural features such as deserts and mountain ranges, or political, showing countries, cities, etc. While most globes emphasize the surface of the land, a globe may also show the bottom of the sea. Globes also can be made to depict the surfaces of spherical bodies other than the Earth, for example, the Moon. Celestial globes are also still in use.

Map

A map is a symbolic representation of selected characteristics of a place, usually drawn on a flat surface. Maps present information about the world in a simple, visual way. They teach about the world by showing sizes and shapes of countries, locations of features, and distances between places. Maps can show distributions of things over Earth, such as settlement patterns. They can show exact locations of houses and streets in a city neighborhood.

Mapmakers, called cartographers, create maps for many different purposes. Vacationers use road maps to plot routes for their trips. Meteorologists—scientists who study weather—use weather maps to prepare forecasts. City planners decide where to put hospitals and parks with the help of maps that show land features and how the land is currently being used. Some common features of maps include scale, symbols, and grids.

Scale

All maps are scale models of reality. A map's scale indicates the relationship between the distances on the map and the actual distances on Earth. This relationship can be expressed by a graphic scale, a verbal scale, or a representative fraction. The most common type of graphic scale looks like a ruler. Also called a bar scale, it is simply a horizontal line marked off in miles, kilometres, or some other unit measuring distance.

The verbal scale is a sentence that relates distance on the map to distance on Earth. For example, a verbal scale might say, "One centimetre represents one kilometre" or "one inch represents eight miles." The representative fraction does not have specific units. It is shown as a fraction or ratio—for example, 1/1,000,000 or 1:1,000,000. This means that any given unit of measure on the map is equal to one million of that unit on Earth. So, 1 centimetre on the map represents 1,000,000 centimetres on Earth, or 10 kilometres. One inch on the map represents 1,000,000 inches on Earth, or a little less than 16 miles.

The size of the area covered helps determine the scale of a map. A map that shows an area in great detail, such as a street map of a neighborhood, is called a large-scale map because objects on the map are relatively large. A map of a larger area, such as a continent or the world, is called a small-scale map because objects on the map are relatively small. Today, maps are often computerized. Many computerized maps allow the viewer to zoom in and out, changing the scale of the map. A person may begin by looking at the map of an entire city that only shows major roads and then zoom in so that every street in a neighborhood is visible.

Symbols

Cartographers use symbols to represent geographic features. For example, black dots represent cities, circled stars represent capital cities, and different sorts of lines represent boundaries, roads, highways, and rivers. Colors are often used as symbols. Green is often used for forests, tan for deserts, and blue for water. A map usually has a legend, or key, that gives the scale of the map and explains what the various symbols represent. Some maps show relief or changes in elevation. A common way to show relief is contour lines, also called topographic lines. These are lines that connect points that have equal elevation. If a map shows a large enough area, contour lines form circles. A group of contour line circles inside one another indicates a change in elevation. As elevation increases, these contour line circles indicate a hill. As elevation decreases, contour line circles indicate a depression in the earth, such as a basin.

Grids

Many maps include a grid pattern, or a series of crossing lines that create squares or rectangles. The grid helps people locate places on the map. On small-scale maps, the grid is often made up of latitude and longitude lines. Latitude lines run east-west around the globe, parallel to the Equator, an imaginary line that circles the middle of the Earth. Longitude lines run north-south, from pole to pole. Latitude and longitude lines are numbered. The intersection of latitude and longitude lines, called coordinates; identify the exact location of a place.

On maps showing greater detail, the grid is often given numbers and letters. The boxes made by the grid may be called A, B, C, and so on across the top of the map, and 1, 2, 3, and so on across the left side. In the map's index, a park's location might be given as B4. The user finds the park by looking in the box where column B and row 4 cross.

Types of Maps

Millions of Unique Maps

Millions of unique maps are in use throughout the world. Most of these maps can be placed into one of two groups:

- Reference maps show the location of geographic boundaries, physical features of Earth, or cultural features such as places, cities, and roads. Political maps, physical maps, road maps, topographic maps, time zone maps, geologic maps, and zip code maps are all examples of reference maps. A variety of reference maps have been created for almost every country of the world.
- Thematic maps show the variation of a topic (the theme) across a geographic area. Weather maps showing daily high temperatures across the United States are familiar examples of a thematic map. They are made by starting with a reference map of the United States. Then temperature data is plotted atop of the reference map using colors to communicate the temperature forecast. Income maps and resource maps are other types of thematic maps.

Political Maps

"Political maps" are among the most widely used reference maps. They are mounted on the walls of classrooms throughout the world. They show the geographic boundaries between governmental units such as countries, states, and counties. They show roads, cities and major water features such as oceans, rivers and lakes. Political maps help people understand the geography of the world. They are usually the first type of map that students are introduced to in school. They are also known as "reference maps" because people refer to them again and again as they have questions.

Political maps are often printed on paper or another physical medium, but they can also be produced in digital form, suitable for viewing online. Every day millions of people visit search engines to find political reference maps. Some of the most popular searches are for "united states map", "world map", "europe map", and "florida map".

Thousands of different political reference maps have been prepared to show the current geography of the United States. There are maps of the entire nation, maps for each of the 50 states, maps of the 3142 counties (parishes in Louisiana, and boroughs in Alaska) that make up the states. Most counties, boroughs and parishes are further subdivided into even smaller political units. An incredible number of political maps have been prepared just to display the geography of the United States. The maps most commonly seen in classrooms and offices are political maps of the world, countries and continents. They are often annotated with push pins, sticky notes, photographs, marker flags and string to show the travels of a family, locations of a business, or other locations and activities worthy of display.

Election Results Maps

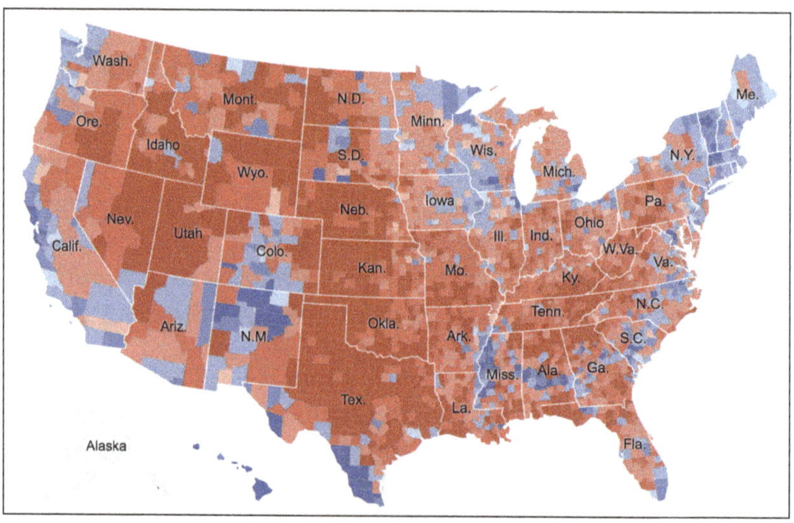

Election Results Map: Sometimes considered to be a different variety of "political map", "election results maps" show the results of an election by geographic subdivision or voting district. The most famous of election maps are the red-state/blue-state maps of the United States presidential election. In recent years they show states won by Republican candidates in red and states won by Democratic candidates in blue. The example above shows the results of the 2016 Presidential election of the United States.

Election results maps might be considered to be a variety of "political maps". These maps show the geographic areas where a candidate for public office received a majority of support from voters. The geographic areas are usually political subdivisions of a country (states), of a state (counties), etc. The most famous examples of election results maps are the red-state/blue-state maps of the United States presidential election. States won by a majority of votes for the Republican candidate are known as "red states", and those won by a majority of votes for the Democratic candidate are known as "blue states". The accompanying map is an example. It maps the results of the 2016 United States presidential election. These maps might be considered "thematic maps" while the election is in progress and while the results are in the news. However, shortly thereafter they might be considered as "reference maps" of historical significance.

Physical Maps

Physical maps are designed to show the natural landscape features of Earth. They are best known for showing topography, either by colors or as shaded relief. Physical maps often have a green to brown to gray color scheme for showing the elevation of the land. Darker greens are used for near-sea-level elevations, with the color grading into tans and browns as elevations increase. The color gradient often terminates in shades of gray for the highest elevations.

Rivers, lakes, seas and oceans are usually shown in blue, often with a light blue color for the shallowest areas and darkening in a gradient or by intervals for areas of deeper water. Glaciers and ice caps are shown in white colors. Physical maps usually show the most important political boundaries, such as state and country boundaries. Major cities and major roads are often shown. This cultural information is not the focus of a physical map, but it is often included for geographic reference and to increase the utility of the map for many users.

Physical Map: This physical map of Eurasia shows the topography of the land in a color-gradient relief. Dark greens are used for near-sea-level elevations, and the green grades to tan and brown as elevation increases. The highest elevations are shown in shades of gray. If you are familiar with the physical features of Eurasia, you can probably recognize the Himalaya Mountain Range, the Tibetan Plateau, the Alps, and the more subtle Ural Mountains. Lake Baikal, the world's largest freshwater lake by volume, can be seen in central Asia.

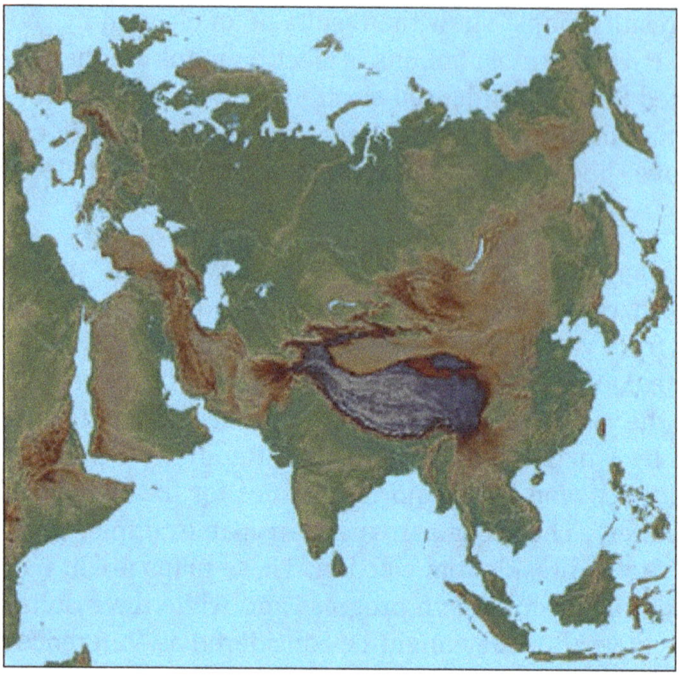

Road, Street and Highway Maps

Google Maps also has special tools that enable you to query "nearby" restaurants, hotels, bars and pubs, museums, pizza, bike shops, schools, attorneys, etc., and the map will populate itself with icons showing their location. If you allow Google Maps to use your current GPS location, you can use the "add destination" tool to plot a car, walking, bike, or public transportation route. Google Maps will even estimate the amount of time required for your trip.

The digital mapping revolution caused an explosion of map creation in the 1990s. In 1996, MapQuest, the first popular online mapping service, allowed anyone with internet access the ability to create customized maps of almost any location in the United States. Within a few months, millions of people had become "cartographers". They were soon producing more unique maps in a single day than had been created during the entire history of paper cartography.

Today, Google Maps is the world's most popular online mapping system. In addition to maps, the service also provides travel route directions. It can create directions for people who are driving, taking public transportation, walking, cycling or taking a plane. Billions of unique maps, millions of travel routes, and millions of street views are created each day with Google Maps. It is the first place millions of people goes to plan any type of travel.

Google has another product named "Google Earth" that allows people to view streets, roads and satellite images within a single interface. Google Earth is a free download - the software installs on your computer and fetches the image directly from the Google

Earth server. Finally, for people who want printed maps, the DeLorme Atlas & Gazetteer is a series of books that contain state-wide map coverage for individual states (or pairs of small adjacent states). The maps present a combination of road, topography, cultural and recreational information. These "hybrid maps" are a favorite of people who work and play outdoors in rural areas.

Topographic Maps

Topographic maps are reference maps that show the shape of Earth's surface. They usually do this with lines of equal elevation known as "contour lines", but elevation can also be shown using colors (second map), color gradients, shaded relief and a number of other methods. Topographic maps are frequently used by hunters, hikers, skiers, and others seeking outdoor recreation. They are also essential tools of the trade for geologists, surveyors, engineers, construction workers, landscape planners, architects, biologists and many other professions - especially people in the military.

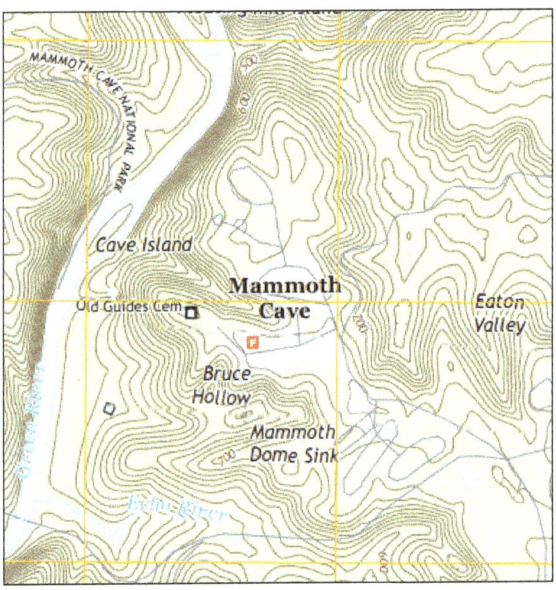

Topographic Map of an area within Mammoth Cave National Park, Kentucky. This map shows Earth's topography using brown contour lines with a contour interval of 20 feet. Roads, place names, streams and other features are also shown. Areas on the map where the brown contour lines are close together have steep slopes. Areas where the contour lines are spaced far apart have gentler slopes.

Topographic maps also show other important natural features such as lakes, rivers and streams. Their locations are determined by topography, making them important natural elements of topographic maps. Important cultural features are also shown on topographic maps. These include roads, trails, buildings, place names, bench marks, cemeteries, churches, schools and much more. A standardized set of special symbols has been developed for this use.

Topographic maps have traditionally been printed on large sheets of paper with their four boundaries being lines of longitude and latitude. The United States Geological Survey is the most widely known organization for producing them. They produce a series of 7.5-minute topographic maps covering most areas of the United States (a 7.5-minute map shows an area that is 7.5 minutes of longitude by 7.5 minutes of latitude). These maps and maps of many other scales are available from USGS in both print and digital form. Commercial publishers of topographic maps include the DeLorme Atlas (paper maps in books with state-wide coverage) and MyTopo (a source of digital and paper maps in traditional topographic and topophoto formats - we are affiliates of MyTopo and receive a commission on referred sales).

Time Zone Maps

Time zones are regions of the world where people set their clocks to display the same time of day. This synchronization of time has many commercial, navigational, and social benefits. By international agreement there are 24 time zones around the world. These 24 zones are shown in the accompanying time zone map. In each of these zones, 12:00 noon occurs at the approximate time of the solar mid-day. The actual solar noon occurs a little earlier in the east side of the time zone and a little later in the west. This variation is caused by Earth's rotation.

Time zone maps are reference maps that people use to determine the time in different parts of the world. For example if a person in New York City wants to phone a person in Los Angeles, he can look at a time zone map and determine that New York City time is three hours ahead of Los Angeles time. This helps people to avoid placing calls outside of business hours and helps people in different time zones schedule meetings and phone calls at mutually agreeable times. Time zones are usually superimposed on a political map of the world or a map of a large single country such as the United States.

Geologic Maps

Geologic maps show the types of rocks and sediments present immediately below the surface of a geographic area. Sediment cover is shown in shades of yellow, and rock units are shown in a variety of colors, often based upon their lithology. Rock unit contacts, faults, folds, and strike and dip measurements are plotted in black.

Geologic maps are important data sources for many types of work. Certain types of rock are used for construction materials, and a geologic map shows where they are located at the surface. Other types of rock might contain valuable minerals, and a geologic map can be used as a preliminary tool for deciding where to drill or prospect. Areas near volcanoes might be underlain by lava flows, lahar deposits, pyroclastic flows or other volcanic products. Geologic maps can be used to conduct a preliminary volcanic hazard assessment of an area. Construction projects require good foundation materials and sources of crushed stone and other materials. Geologic maps can be used for the preliminary work of identifying potentially stable sites near economic sources of suitable construction materials.

Geologic maps are made by geologists in the field who identify, sample, and measure the rocks. Because rocks are not exposed in all locations - especially in areas with heavy vegetation - they are often based upon fragmentary information. This fragmentary information can be supplemented when construction projects, landslides, stream erosion and other events expose rock beneath previously unobserved areas. As a result, geologic maps can be refined and updated as new information is obtained.

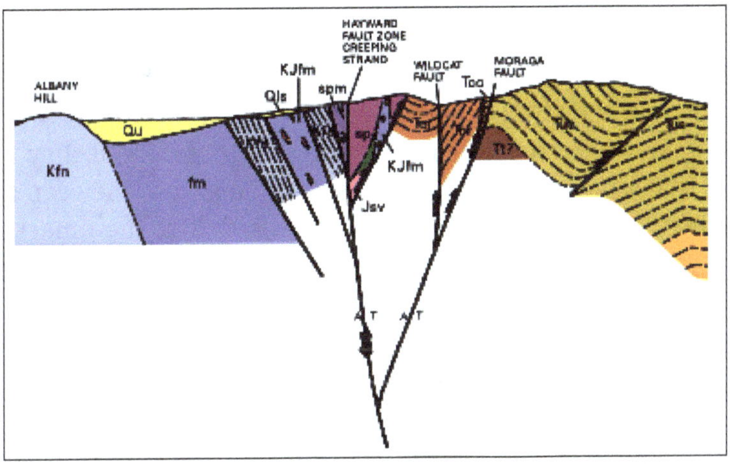

Geologic Cross-Section representing the subsurface structure of rocks in the map area above. This cross-section represents a southwest-northeast slice through the map area, showing folds, faults, a sediment lens and a view of the creeping portion of the Hayward Fault.

Most geologic maps are usually accompanied by at least one cross-section that illustrates what is expected to be seen if a "slice through the Earth" was cut across the map area. These cross-sections illustrate the geologic structures inferred by mapping the rocks and sediments above.

Zip Code Maps

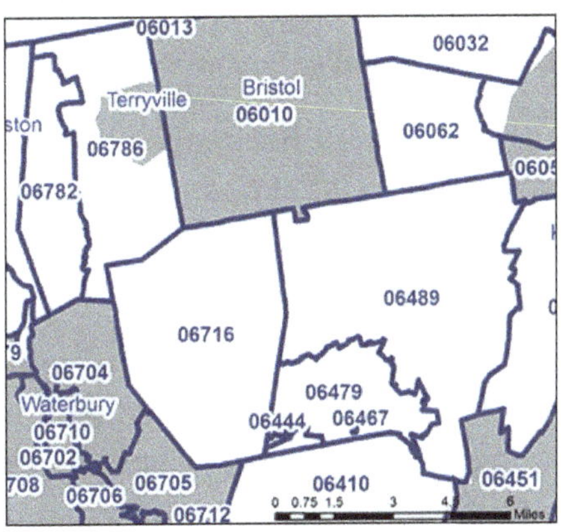

Zip Code Map: This map shows some of the zip codes in the state of Connecticut.

Zip Code Maps are maps that show the approximate boundary of zip code areas used by the United States Postal Service. They are usually plotted over a base map that shows the roads and streets within the zip code area. The United States Postal Service assigns zip codes for a section of a street, a collection of streets, an establishment, a structure, a group of post office boxes, or the area serviced by a single post office for the delivery of mail. Rather than consisting of geographic areas, zip codes are more in conformance with a group of mail delivery routes. In sparsely populated areas, a single zip code can cover many square miles, but in cities a zip code can be assigned to a single building or to an organization with a campus of buildings.

Businesses make valuable use of zip code maps by matching them to zip code tabulation data compiled by the United States Census Bureau. This data characterizes the population within a zip code by age, gender, race, national origin, income, housing, and much more. Companies can use this information to determine if they want to market to people within that zip code and how they want to market. They can also coordinate their mailings with the United States Postal Service to deliver their marketing materials to zip codes where demographic data suggests a high density of potential customers.

Weather Maps

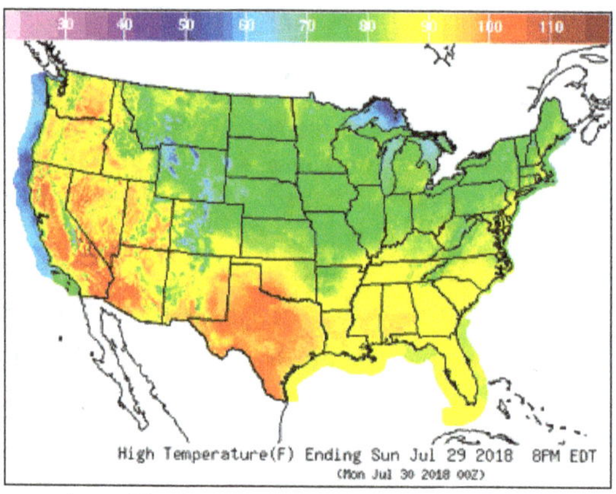

Weather Map showing projected high temperatures. This is one of the many types of weather maps prepared by the National Weather Service of the United States and published online for anyone to use.

People use an incredible number of weather maps. They are used to show predicted temperatures, predicted precipitation, storm warnings of various kinds, wind speed and direction, chance of precipitation, type of precipitation, snow accumulation, frost prediction and many other aspects of weather.

All of these weather maps are continuously updated to communicate the most current information. They are the world's most frequently consulted thematic maps. Weather maps are presented in newspapers, television programs and especially on websites. Delivering weather maps on websites and through web apps gives people around the world instant access to weather information. Many weather maps are animated maps that show historical or projected changes in the weather. These are extremely useful for people who need to know how changes in weather will impact their travel, workday, recreation, and many other plans.

Species Distribution Maps

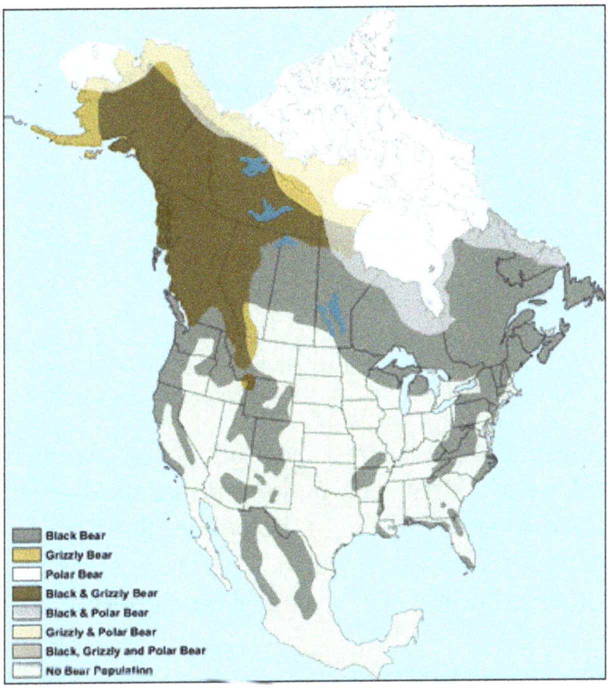

Species Distribution Maps show where certain plant or animal species can be found.

The area normally inhabited by a plant or animal species can be mapped. These maps are made by compiling the locations where qualified people have repeatedly observed that species in the wild. The presence / absence of the species define its geographic range. The resulting maps show (by color or pattern) the geographic area where that species is encountered. These maps are known as "species distribution maps" or "geographic range maps." A website wanted a map that displayed the geographic distribution of bears in North America. That map was made by overlaying the geographic ranges of black bears, grizzly bears, and polar bears on a single map. Because the geographic ranges of these three species overlap, it is possible to be in an area where a person might encounter one, two, or even three different species. There are also areas of North America where a person would not expect to encounter a bear. The resulting map is shown in the accompanying image.

Income Maps

Income maps are a very common type of thematic map. They show variations of income across a geographic area. The standard mapped variable for an income map is median household income. Income tends to be highly geographic because rural portions of a state or country often have lower median household income than urban areas. Within urban areas, income can also be highly variable because neighborhoods tend to be populated by people with similar income levels.

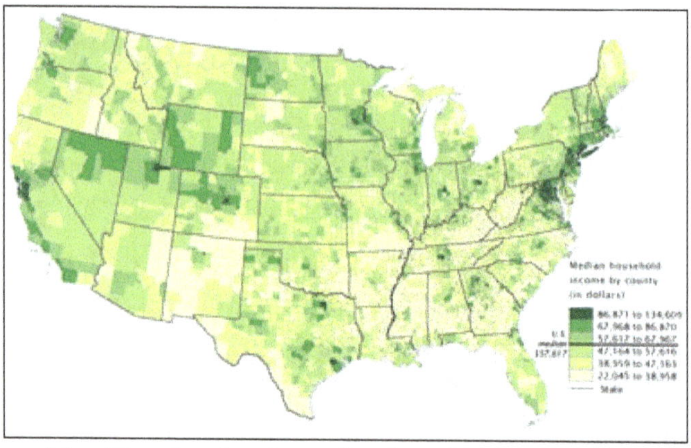

Income Map: Map showing the median household income of the United States using individual counties as enumeration units.

The United States Census Bureau is a regular producer of income maps for the United States and individual states. After each major census, the Bureau updates its set of income maps and makes them available to the public. The Census Bureau also makes "change in income maps". These show which geographic areas have experienced economic growth or economic decline over a specific time interval.

Resource Maps

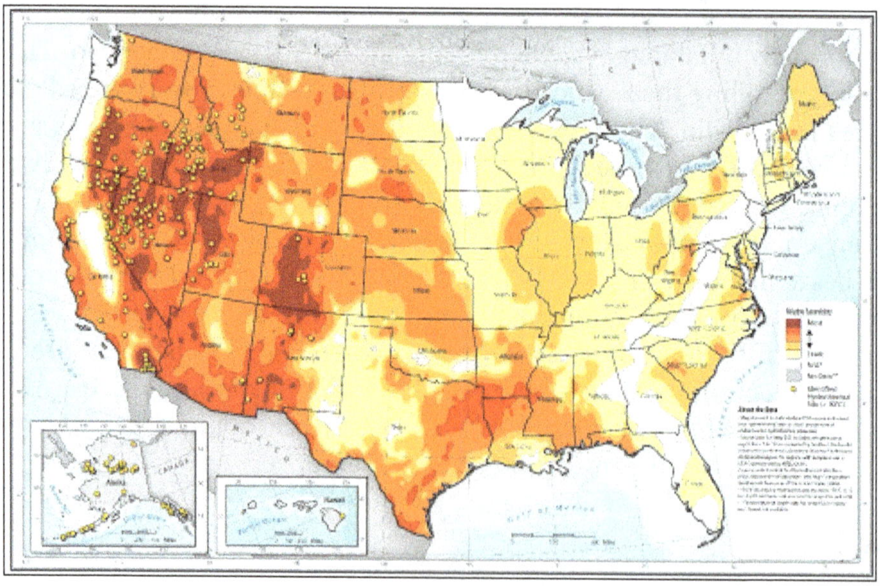

Resource Map: Map illustrating the photovoltaic solar resource of the United States. It is clear from this map that the solar resource of the United States is greatest in the south western part of the country. Citizens, companies and governments can use this map to make decisions about investing in a solar power solution.

Thematic maps are often made to communicate the geographic distribution of natural resources. These maps might show countries with the highest diamond production or the geographic extent of an oil or gas field. The map shown here illustrates the geographic pattern of solar generating capacity for the United States.

Resource maps are important because they help governments understand their natural resource assets and the natural resource assets of their allies and potential enemies. Resource maps help mining companies target their exploration efforts. They are also important for assessing the transportation opportunities and problems associated with the distribution of resources and the location of where they are consumed.

Earthquake Maps

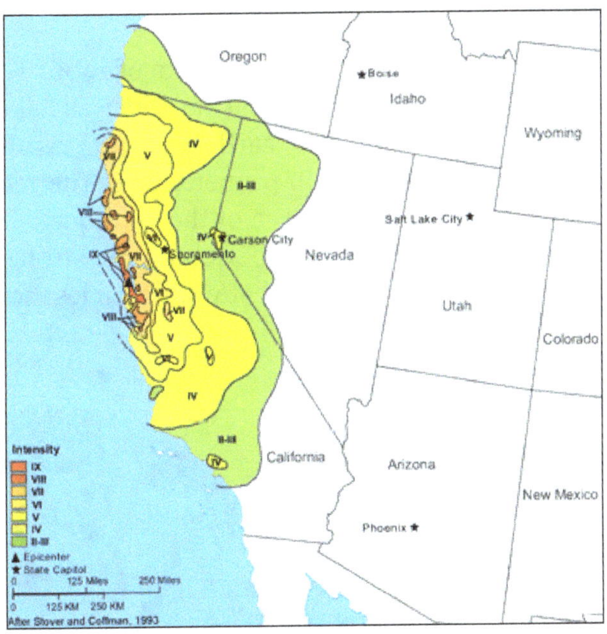

Earthquake maps focus on the scientific aspects of earthquakes or on the impact and mitigation aspects. There are many types of earthquake maps. Here are a few:

- Epicenter Maps show the geographic location of earthquake epicenters (the point on Earth's surface immediately above where the vibrations of an earthquake originated). They can show the location of one earthquake or the geographic distribution of many earthquakes. They have scientific value for understanding the distribution of earthquakes and mitigation value because they reveal areas where earthquakes frequently occur. Each epicenter is plotted as a symbol, and the symbols can have different colors, sizes and shapes to indicate the earthquakes' strength, depth, or date.

- Fault Maps show the geographic location of faults. Faults are fractures in the Earth from which earthquakes and their vibrations emanate. Fault maps are of interest to scientists, and they are a geologic hazard map because future movements of the fault will generate the vibrations of an earthquake.
- Intensity Maps show the geographic distribution of ground shaking caused by an earthquake. The intensity of shaking is measured using the Modified Mercalli Intensity Scale. Intensity maps are produced by obtaining intensity values from many locations within the felt area of the earthquake, plotting that data on a map, and contouring that data into areas of equal intensity. Intensity maps can be used as a measure of future earthquake hazard.
- Earthquake Hazard Maps show geographic areas subject to specific types of hazards. They might show areas prone to the liquefaction of soil, areas prone to land sliding, and areas underlain by soft sediment that might cause unusually intense vibrations.

Intensity Map for the Great San Francisco Earthquake of 1906: This is the most devastating earthquake in California history. The earthquake was felt throughout most of California and into Nevada and Oregon. The earthquake and resulting fires caused an estimated 3000 deaths and $524 million in property loss. The map above shows the intensity of ground shaking across the felt area of the earthquake. Intensity levels are shown as a color sequence where red areas experienced the strongest ground shaking and green areas experienced ground shaking that was much less severe.

Plate Tectonics Maps

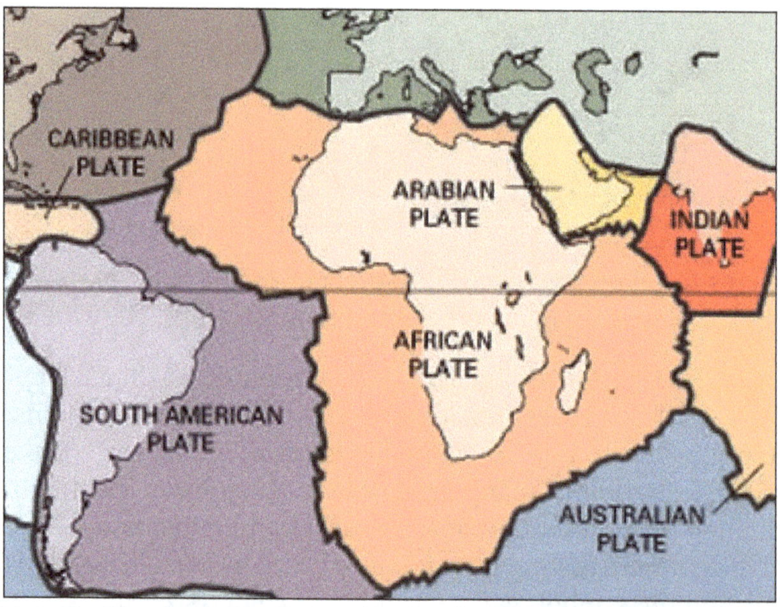

Plate Tectonics Map: A portion of a map showing the geographic extent and boundaries of Earth's lithospheric plates by the United States Geological Survey.

Plate tectonics is a branch of the geological sciences that deals with the movement and interaction of Earth's lithospheric plates. These plates move and interact with one another to cause earthquakes, volcanic activity, mountain building, the production of new lithosphere, the destruction of old lithosphere, and other geologic processes. Most of this activity occurs at lithospheric plate boundaries.

Geologists produce and use many types of maps in the study of plate tectonics. The most commonly referenced is a plate boundary map. These maps show the geographic extent and boundaries of the lithospheric plates. Other types of maps used in the study of plate tectonics include: earthquake epicenter maps, volcanic activity maps, ocean floor age maps, plate movement direction maps, and many others. Much of the work done by geologists involves the production of a map.

Bathymetric Maps

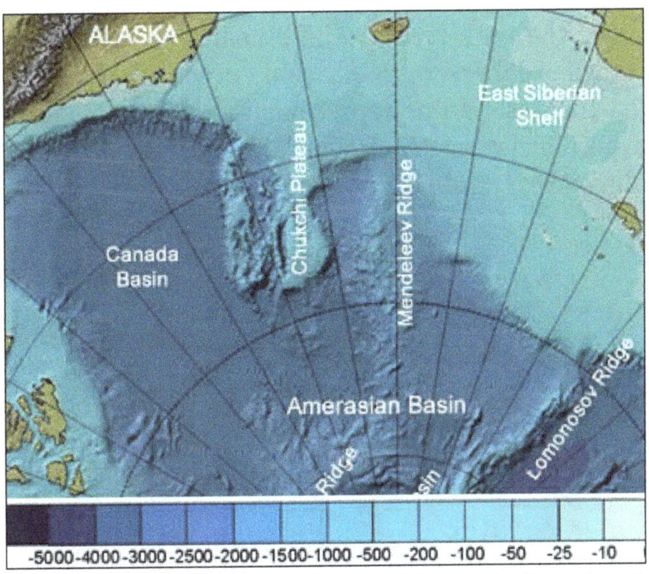

Bathymetric Map: A portion of a bathymetric map of the Arctic Ocean in color and shaded relief. Map by Intergovernmental Oceanographic Commission and others.

A bathymetric map, sometimes called a bathymetric chart, is a map that shows the depth of a body of water, such as a lake, a part of the ocean, a river, or a reservoir. Depths can be shown using contour lines, shaded relief, colors, or a combination of these methods. Bathymetric maps are valuable for avoiding shallow or dangerous waters during ship navigation, planning construction projects, deciding where to fish, determining the cost of drilling for oil in the Gulf of Mexico, planning a trans-Atlantic communications cable, or finding a good place to scuba dive. Making bathymetric maps used to be very expensive and time consuming. Depths were determined by "soundings" - lowering a weight on a cable to the bottom of the water body and determining the length of cable required. Today, depths can be determined by sonar and a variety of other remote measurement methods.

Volcanic Hazard Maps

Did you know that scientists have identified 161 volcanoes in the United States that will most likely erupt at some time in the future? Most of these volcanoes are in Alaska, but significant numbers are found in Hawaii, California, Washington, and Oregon. People need to know which volcanoes present a danger, the types of volcanic hazards that can be expected, and the geographic area that will be dangerous if an eruption occurs. For that reason, geologists have prepared geologic hazard maps for the volcanoes that present the greatest risk to people and are working to complete maps for other volcanoes.

The accompanying map shows three areas designated to have a high to very high volcanic threat in northern California: Medicine Lake Volcano, Mount Shasta, and Lassen Volcanic Center. At Medicine Lake Volcano: the dashed line represents the area that could potentially receive ashfalls of two inches or greater; the yellow area is where lava flows could occur; and the pink area is where near-vent hazards such as pyroclastic flows, lava flows, ballistics, lahars, or floods could occur. At Mount Shasta and the Lassen Volcanic Center, the dashed line represents the two-inch ashfall hazard area, the pink areas could experience near-vent hazards, the brown areas could experience lahars, and the dark blue areas could experience floods.

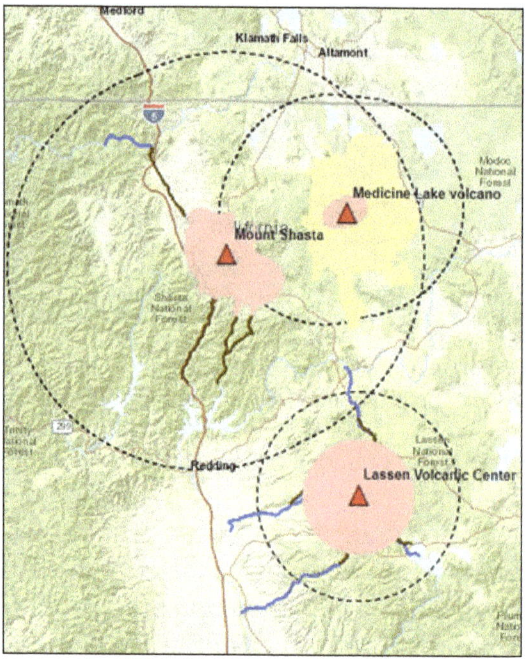

Volcanic Hazard Map for three volcanic areas: Medicine Lake Volcano, Mount Shasta, and Lassen Volcanic Center.

Hurricane Track Maps

Hurricanes and tropical storms can cause enormous amounts of damage. When one of these storms is detected, the National Hurricane Center starts monitoring it and determining its anticipated path.

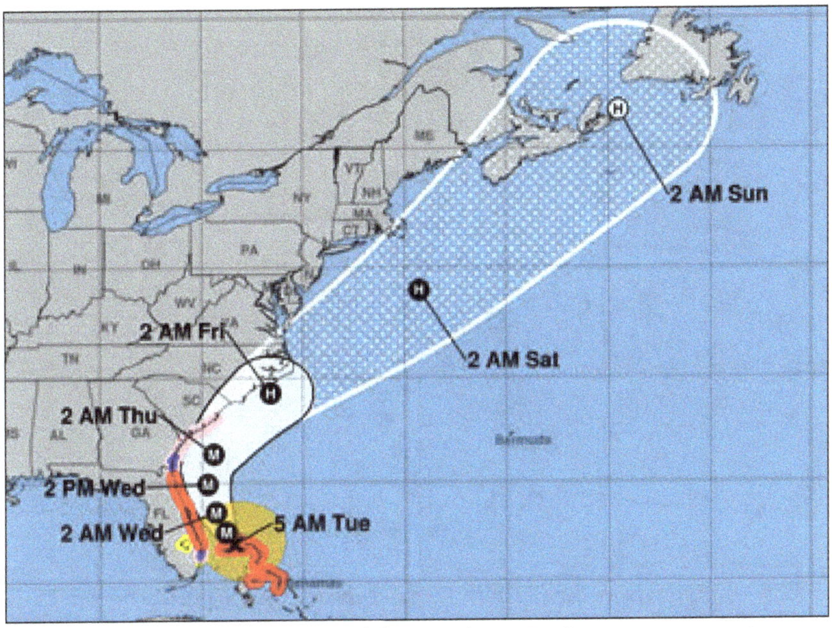

A Hurricane Track Map prepared for Hurricane Dorian, a category 5 storms that made landfall on the Florida, Georgia, and Carolinas coasts.

Map Projection

Map projections are logical and methodical transformations that permit the orderly representation of the Earth's spherical graticule on a flat map. Mathematically speaking, map projections are transformations of geographic co-ordinates (latitude, longitude) into the Cartesian (x, y) co-ordinate space of the map. A map projection is the manner in which the three dimensional spherical surface of the Earth is represented on a two-dimensional surface of a piece of paper. The term 'projection' comes from the idea of placing a light source within a transparent globe and projecting shadows of the meridians, and parallels onto a sheet of paper placed tangent to the globe. It can be defined as a "systematic arrangement of intersecting lines on a plane that represent and have one to one correspondence with the meridians and parallels on the datum surface". Each map projection has certain strengths and weaknesses in terms of accuracy of shape, area, distance and direction. It is not possible for any of the projection to retain more than one of these characteristics over a large part of the earth. It is important to identify that because of the curvature of the earth, all map projections distort distance and directional relationships. The effect of curvature on the distances and directions can be safely ignored for a small area but for a larger area distortion cannot be avoided.

Classification of Map Projection

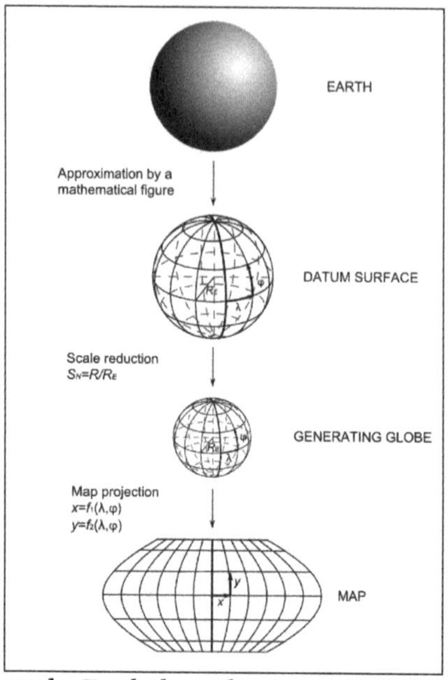

Map projection from the Earth through a generating globe to the final map.

The process of map projection is accomplished in three specific steps:

- Approximating the size and shape of the object (e.g., earth), by a mathematical figure that is by a sphere or an ellipsoid.
- Reducing the scale of the mathematical representation to a generating globe (a reduced model of the earth from which map projections are made) with the principal scale or nominal scale that is the ratio of the radius of the generating globe to the radius of the mathematical figure representing the object [earth]) equivalent to the scale of the plane map.
- Transforming the generating globe into the map using a map projection.

Map projections depend first on an assumption of specific parameters of the object (Earth) itself, such as spherical or ellipsoidal shape, radius of the sphere (or lengths of the semi-major and semi-minor axes of the ellipsoid), and a specific datum or starting point for a coordinate system representation. These assumptions form the basis of the science of Geodesy and are currently accomplished using satellite measurements usually from the Global Positioning System (GPS), Glonass, or Galileo. Once these measurements are accepted, an ellipsoidal representation of coordinates is generated as latitude and longitude coordinates. Those coordinates can then be transformed through map projection equations to a plane Cartesian system of x and y coordinates. The general equations of these transformations have the following form:

$$x = f_1(\phi, \lambda), y = f_2(\phi, \lambda)$$

where,

x is the plane coordinate in the east-west direction,

y is the plane coordinate in the north-south direction,

ϕ is the latitude coordinate,

λ is the longitude coordinate.

The form of the functions f_1 and f_2 determines the exact transformation and the characteristics of the ellipsoidal or spherical representation that will be preserved. Before addressing the specific types of transformations and the characteristics preserved, it is necessary to understand the geodetic characteristics of the ellipsoidal coordinates and how these are generated with modern satellite positioning systems.

Geodesy and Global Navigation Satellite Systems (GNSS)

Map projections have their largest and most frequent application in producing maps showing a smaller or bigger part of the Earth's surface. In order to produce the map of a region, it is necessary to make a geodetic survey of that region and then to visualise the results of such a survey. Geodesy is a technology and science dealing with the survey and representation of the Earth's surface, the determination of the Earth's shape and dimensions and its gravity field. Geodesy can be divided into applied, physical, and satellite geodesy.

Applied geodesy is a part of geodesy encompassing land surveying, engineering geodesy and management of geospatial information. Land surveying is a technique for assessing the relative position of objects on the Earth surface, when the Earth's curvature is not taken into account. Engineering geodesy is a part of geodesy dealing with designing, measuring, and supervising of constructions and other objects (e.g., roads, tunnels and bridges).

Physical geodesy is a part of geodesy dealing with the Earth's gravity field and its implication on geodetic measurements. The main goal of physical geodesy is the determination of the dimensions of the geoid, a level surface modelling Earth, where the potential of the gravity field is constant. Geometrical geodesy is concerned with determination of the Earth's shape, size, and precise location of its parts, including accounting for the Earth's curvature.

Satellite geodesy is part of geodesy where satellites are used for measurements. In the past, exact positions of isolated spots on the Earth were determined in astronomical geodesy, that is, by taking measurements on the stars. Measuring techniques in satellite geodesy are geodetic usage of Global Navigation Satellite Systems (GNSS) such as GPS, Glonass and Galileo.

A satellite navigation system is a system of satellites that provides autonomous geo-spatial positioning with global coverage. It allows small electronic receivers to determine their location (longitude, latitude, and altitude) to within a few metres using time signals transmitted along a line-of-sight by radio from satellites. Receivers calculate the precise time as well as position. A satellite navigation system with global coverage may be termed a global navigation satellite system or GNSS. As of April, only the United States NAVSTAR Global Positioning System (GPS) and the Russian GLONASS are global operational GNSSs. China is in the process of expanding its regional Beidou navigation system into a GNSS by 2020. The European Union's Galileo positioning system is a GNSS in initial deployment phase, scheduled to be fully operational by 2020 at the earliest. France, India and Japan are in the process of developing regional navigation systems. Global coverage for each system is generally achieved by a satellite constellation of 20–30 medium Earth orbit satellites spread among several orbital planes. The actual systems vary but use orbital inclinations of >50° and orbital periods of roughly twelve hours at an altitude of about 20,000 kilometres.

Photogrammetry is an important technology for acquiring reliable quantitative information on physical objects and the environment by using recording, measurements and interpretation of photographs and scenes of electromagnetic radiation by using sensor systems. Remote sensing is a method of collecting and interpreting data of objects from a distance. The method is characterized by the fact that the measuring device is not in contact with the object to be surveyed. Its most frequent application is from aerial or space platforms.

The study of the transformation from the Earth's surface model or generating globe to a two-dimensional representation requires the use of the following concepts: ellipsoid, datum, and coordinate system. The Earth's ellipsoid is any ellipsoid approximating the Earth's figure. Generally, an ellipsoid has three different axes, but in geodesy and cartography, it is most often a rotational ellipsoid with small flattening.

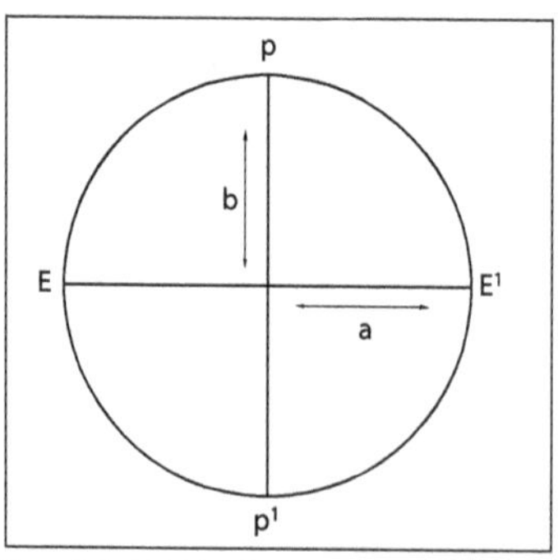

Terminology for rotational ellipsoid: EE' is the major axis, PP' is the minor axis and the axis of rotation, where a, is the semi-major axis and b is the semi-minor axis.

The rotational ellipsoid is a surface resulting from rotating an ellipse around a straight line passing through the endpoints of the ellipse. It is used to model the Earth. Famous

Earth ellipsoids include the ones elaborated by Bessel, and the more recently, WGS84 and GRS80 ellipsoids. Flattening is a parameter used to determine the difference between the ellipsoid and the sphere. It is defined by the equation $f = \frac{a-b}{a}$, where a and b are the semi-major and semi-minor axes, respectively. The semi- major axis a, is the Equatorial radius because the Equator is a circle. The semi-minor axis b is not a radius, because any planar section of the ellipsoid having poles P and P' as common points is an ellipse and not a circle.

Generally speaking, a datum is a set of basic parameters which are references to define other parameters. A geodetic datum describes the relation of origin and orientation of axes on a coordinate system in relation to Earth. At least eight parameters are needed to define a global datum: three for determination of the origin, three for the determination of the coordinate system orientation and two for determination of the geodetic ellipsoid. A two-dimensional datum is a reference for defining two-dimensional coordinates on a surface. The surface can be an ellipsoid, a sphere or even a plane when the region of interest is relatively small. A one-dimensional datum or vertical datum is a basis for definition of heights and usually in some relation to mean sea level.

The WGS84 and GRS80 ellipsoids were established by satellite positioning techniques. They are referenced to the centre mass of the Earth (i.e., geocentric) and provide a reasonable fit to the entire Earth. The WGS84 datum provides the basis of coordinates collected from the GPS, although modern receivers transform the coordinates into almost any user selected reference datum. The need for datum transformation arises when the data belongs to one datum, and there is a need to get them in another one (e.g., WGS84 to North American Datum or vice versa).

Three-Dimensional Coordinate Reference Systems

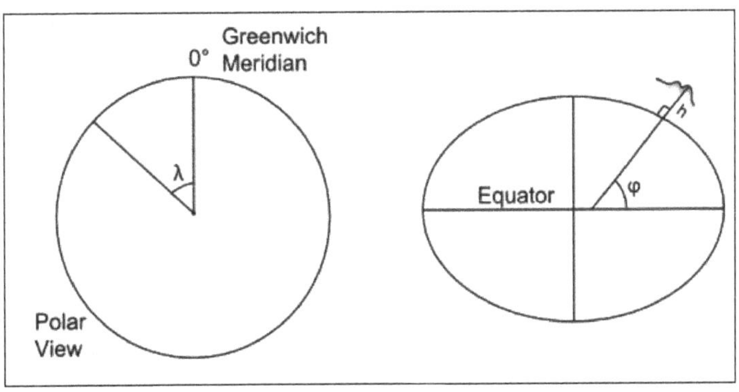

Geodetic or ellipsoidal coordinate system.

Geodetic coordinates are geodetic latitude and geodetic longitude, with or without height. They are also referred to as ellipsoidal coordinates. Geodetic latitude is a parameter which determines the position of parallels on the Earth's ellipsoid and is defined by the angle from the equatorial plane to the normal one (or line perpendicular) to the ellipsoid at a given point. It is usually from the interval [−90°, 90°] and is marked with Greek letter φ. An increase in geodetic latitude marks the direction of North, while its decrease marks the direction south. Geodetic longitude is a parameter which determines the position of the meridian on the Earth's ellipsoid and is defined by the angle from the prime meridian (that is the meridian of the Greenwich observatory near London) plane to the given point on the meridian plane. It is most often from the interval [−180°, 180°] and is marked with Greek letter λ. An increase in geodetic longitudes determines the direction of East, while a decrease determines the direction of West.

A geodetic datum should define the relation of geodetic coordinates to the Earth. Geodetic coordinates φ, λ and height h may be transformed to an Earth-centred, Cartesian three-dimensional system using the following equations:

$$X = (N + h)\cos\varphi\cos\lambda$$
$$Y = (N + h)\cos\varphi\sin\lambda$$
$$Z = \left(N\left(1 - e^2\right) + h\right)\sin\varphi$$

where

$$N = \frac{a}{\sqrt{1 - e^2\sin^2\varphi}}, e^2 = \frac{a^2 - b^2}{a^2}$$

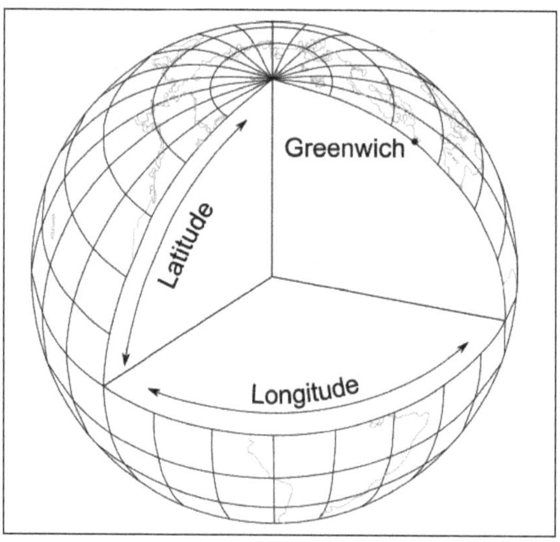

Geographic or spherical coordinate system: geographic latitude φ, geographic longitude λ.

If we wish to represent a large part of the Earth, a continent or even the whole world, the flattening of the Earth can be neglected. In that case, we speak about a geographic coordinate system instead of a geodetic coordinate system. Geographic coordinates are geographic latitude and geographic longitude, with or without height. They are also referred to as spherical coordinates. Geographic latitude is a parameter which determines the position of parallels on the Earth's sphere and is defined by the angle from the equatorial plane to the normal on the sphere at a given point. It is usually from the interval [−90°, 90°] and is marked with Greek letter φ. An increase in geographic latitude marks the direction of North, while its decrease marks the direction south. Geographic longitude is a parameter which determines the position of the meridian on the Earth's sphere and is defined by the angle from the prime meridian plane to the given point on the meridian plane. It is most often from the interval [−180°, 180°] and is marked with Greek letter λ. An increase in geographic longitudes determines the direction of East, while a decrease determines the direction of West.

Geographic coordinates φ, λ and height h = 0 may be transformed to an Earth-centred, Cartesian three-dimensional system using the following equations:

$$X = R\cos\varphi\cos\lambda$$
$$Y = R\cos\varphi\sin\lambda$$
$$Z = R\sin\varphi$$

where R is a radius of the spherical Earth.

A spherical coordinate system can be obtained as a special case of an ellipsoidal coordinate system taking into account that flattening equals zero, f = 0, or equivalently stating that the second eccentricity equals zero, e = 0. Sometimes, in geodetic and cartographic practice, it is necessary to transform Cartesian three-dimensional coordinates to spherical or even ellipsoidal coordinates. Furthermore, sometimes there is a need to make a transformation from one three-dimensional coordinate system to another one.

Two-Dimensional Coordinate Reference Systems

Generally, for use of geospatial data, a common frame of reference is needed and this is usually done in a plane reference system. Because maps reside in a plane geometric system, the spherical or ellipsoidal coordinates, generated from satellite positioning systems or from any other surveying device, must be mathematically transformed to the plane geometry system. The simplest transformation is to assume that the plane x coordinate is equivalent to φ, and the plane y coordinate is equivalent to λ. The result is known as the Plate Carrée projection and although it is simple, it involves significant distortion of the coordinate positions and thus presents areas, most distances, and angles that are distorted or deformed in the plane.

More sophisticated transformations allow preservation of accurate representations of area or distance or angles, or other characteristics, but not all can be preserved in the same transformation. In fact, usually only a single characteristic, for example preservation of accurate representation of area, can be maintained, resulting in distortion of the other characteristics. Thus, many different map projections have been developed to allow preservation of the specific characteristics a map user may require.

The Universal Transverse Mercator (UTM) coordinate system is based on projections of six-degree zones of longitude, 80° S to 84° N latitude and the scale factor 0.9996 is specified for the central meridian for each UTM zone yielding a maximum error of 1 part in 2,500. In the northern hemisphere, the x coordinate of the central meridian is offset to have a value of 500,000 meters instead of zero, normally termed as "False Easting." The y coordinate is set to zero at the Equator. In the southern hemisphere, the False Easting is also 500,000 meters with a y offset of the Equator or False Northing equal to 10,000,000 meters. These offsets force all coordinates in the system to be positive. In the Universal Military Grid System (UMGS), the polar areas, north of 84 °N and south of 80 °S, are projected to the Universal Polar Stereographic (UPS) Grid with the pole as the centre of projection and a scale factor 0.9994. They are termed "North Zone" and "South Zone."

Map projection also is dependent on the shape of the country. In the United States of America, the State Plane Coordinate System is established in which states with an east-west long axis, Tennessee; for example, use the Lambert Conformal Conic projection, whereas states with a north-south long axis, Illinois, for example, use the Transverse Mercator projection. Not only a map projection and the map scale, but coordinate measurement units are also an important part of any map. In order to be sure of the accuracy of data taken from a map, read carefully all information written along the border of the map and, if necessary, ask the National Mapping Agency for additional information.

A final plane coordinate system of relevance to geographic data modelling and analyses, particularly for satellite images and photographs, is an image coordinate system. A digital image system is not a right-handed Cartesian coordinate system since usually the initial point (0, 0) is assigned to the upper left corner of an image. The x coordinate, often called sample, increases to the right, but the y coordinate, called the line, increases down. Units commonly are expressed in picture elements or pixels. A pixel is a discrete unit of the Earth's surface, usually square with a defined size, often expressed in metres.

Often, in geodetic and cartographic practice, it is necessary to transform plane Cartesian two-dimensional coordinates to another plane two-dimensional coordinate system. The indirect method transforms plane two-dimensional coordinates into spherical or ellipsoidal coordinates by using so-called inverse map projection equations. Then, the method follows with appropriate map projection equations that give the result in the second plane, two-dimensional system. The direct method transforms plane coordinates from one system to another by using rotation, translation, scaling, or any other two-dimensional transformation.

Classes of Map Projections

Projections may be classified on the basis of geometry, shape, special properties, projection parameters, and nomenclature. The geometric classification is based on the patterns of the network (the network of parallels of latitude and meridians of longitude). According to this classification, map projections are usually referred to as cylindrical, conical, and azimuthal, but there are also others.

An azimuthal projection also projects the image of the Earth on a plane. A map produced in cylindrical projection can be folded in a cylinder, while a map produced in conical projection can be folded into a cone. Firstly, let us accept that almost all map projections in use are derived by using mathematics, especially its part known as differential calculus. This process allows for the preservation of specific characteristics and minimizing distortion, such as angular relationships (shape) or area.

Cylindrical Projections

Cylindrical projections are those that provide the appearance of a rectangle. The rectangle can be seen as a developed cylindrical surface that can be rolled into a cylinder. Whereas these projections are created mathematically rather than from the cylinder, the final appearance may suggest a cylindrical construction. A cylindrical map projection can have one line or two lines of no scale distortion. Classic examples of cylindrical projections include the conformal Mercator and Lambert's original cylindrical equal area.

Cylindrical projections are often used for world maps with the latitude limited to a reasonable range of degrees south and north to avoid the great distortion of the polar areas by this projection method. The normal aspect Mercator projection is used for nautical charts throughout the world, while its transverse aspect is regularly used for topographic maps and is the projection used for the UTM coordinate system.

a.

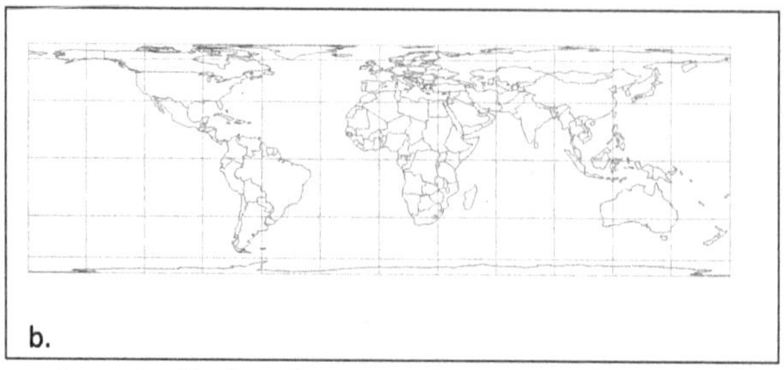

The conformal cylindrical Mercator projection (a) and Lambert's cylindrical equal area projection (b).

Conical Projections

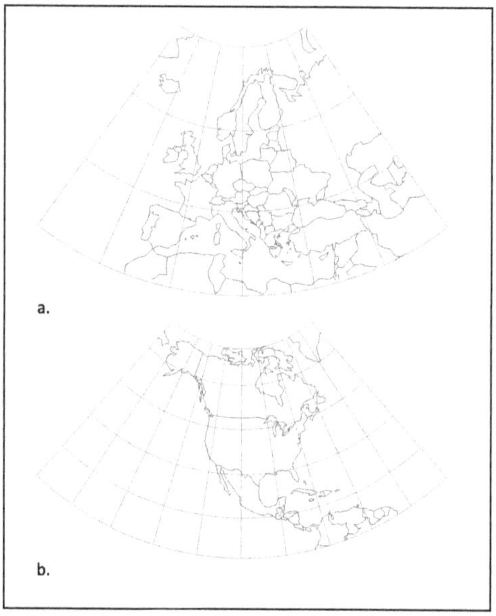

Lambert's conformal conic (a) and the Albers conical equal area (b) projections.

Conical projections give the appearance of a developed cone surface that can be furled into a cone. These projections are usually created mathematically and not by projecting onto a conical surface. A single line or two lines may exist as lines of no scale distortion. Classic examples of conical projections are Lambert's conformal conic and the Albers conical equal area projection. Conical projections are inappropriate for maps of the entire Earth and work best in areas with a long axis in the east-west direction. This makes them ideal for representations of land masses in the northern hemisphere, such as the United States of America, Europe, or Russia.

Azimuthal Projections

Azimuthal projections are those preserving azimuths (i.e., directions related to north in its normal aspect). A single point or a circle may exist with no scale distortion. Classic examples of azimuthal projections include the stereographic and Lambert's azimuthal equal area.

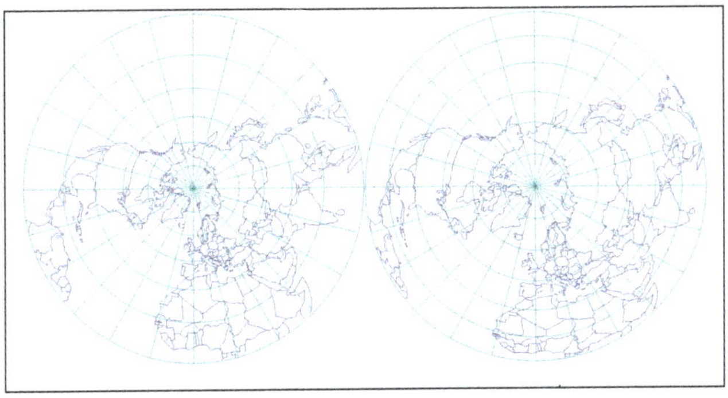

The stereographic (a) and Lambert's azimuthal equal-area (b) projections.

Other classifications of map projections are based on the aspect (i.e., the appearance and position of the graticule, poles or the equator in the projection). Aspect can be polar, equatorial, normal, transverse or oblique. Accordingly, there are polar projections, normal projections, equatorial projections, transverse projections and oblique map projections. These are names of individual sets of map projections and not a systematic categorization because, for example, a projection can be polar and normal at the same time. In theory, each projection can have any aspect. However, many projections are almost always used in certain aspects in order to express their characteristics as well as possible.

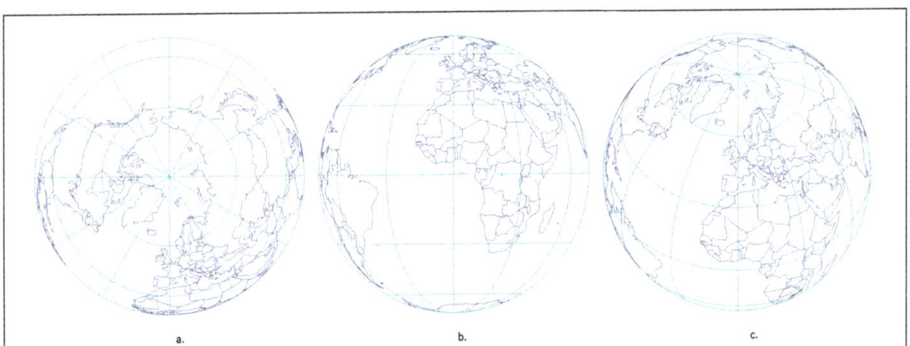

Orthographic projection in its normal (a), transverse (b) and oblique (c) aspects.

For example, many factors such as temperature, contamination breakout and biodiversity depend on the climate (i.e., the latitude). For projections with a constant distance between parallels, the latitude in the equatorial aspect can be directly converted into vertical distance, facilitating comparison. Certain projections with graticule in normal aspect appearing as simple curves were originally defined by geometric constructions.

Considering most transverse and oblique projections have graticule consisting of complex curves, such projections were not systematically analysed prior to the computer era. In general, calculating oblique projections for a particular ellipsoid is very complex and is not developed for all projections. Nevertheless, oblique projections have applications.

A map projection is a normal projection or it is in normal aspect if the appearance and position of the graticule, poles and the equator in the projection are the most natural and are usually determined by geometrical conditions. It is often determined by the simplest calculations or the simplest appearance of the graticule. The polar aspect is normal for azimuthal projections, while the equatorial aspect is normal for cylindrical projections. In azimuthal and conic projections, the graticule consists of straight lines and arcs of circles; normal aspect cylindrical projections have graticule consisting only of straight lines forming a rectangular grid.

A map projection is a transverse projection or it is in transverse aspect if the appearance and position of the graticule, poles or the equator in the projection were derived by applying formulae for the normal aspect projection to a globe which was previously rotated by 90° around its centre, so that poles are in the equatorial plane.

A map projection is a polar projection or it is in polar aspect if the image of a pole is in the centre of the map. It is often used as a synonym for normal aspect azimuthal projection. A map projection is equatorial or it is in equatorial aspect if the image of the equator is in the centre of the map. The image of the Equator is placed in the direction of one of the main axes of the map, mostly horizontally. Equatorial projection often means normal aspect cylindrical projection. A map projection is an oblique projection or it is in oblique aspect if it is neither polar nor equatorial, neither normal aspect nor transverse.

Preserving Specific Properties with Map Projections

Map projections are usually designed to preserve specific characteristics of the globe, such as areas, angles, distances, or specific properties such as great circles (intersections of the Earth and a plane which passes through the Earth's center) becoming straight lines. Maps with angles preserved are called conformal projections. Maps with areas preserved are referred to as equal-area or equivalent projections.

Preserving Angles

Gerardus Mercator in 1569 developed a cylindrical conformal projection that bears his name. He developed it to show loxodromes or rhumb lines, which are lines of constant bearing, as straight lines, making it possible to navigate a constant course based on drawing a rhumb line on the chart. The Mercator projection has meridians as equally spaced parallel lines, with parallels shown as unequally spaced straight parallel lines, closest near the Equator and perpendicular to the meridians. The North and South Poles cannot be shown. Scale is true along the Equator or along two parallels equidistant from the Equator. Significant size distortion occurs in the higher latitudes and that is why the Mercator projection is not recommended for world maps. The Mercator projection, a standard for marine charts, was defined for navigational charts and is best used for navigational purposes.

Transverse Mercator

The transverse Mercator, also known as a Gauss-Krüger projection, is a projection where the line of constant scale is along a meridian rather than the Equator. The central meridian and the Equator are straight lines. Other meridians and parallels are complex curves and are concave toward the central meridian. The projection has true scale along the central meridian or along two lines equidistant from and parallel to the central meridian. It is commonly used for large-scale, small area, presentations. Due to the distribution of distortion, it is usually used by dividing the region to be mapped in three-degree or six-degree zones limited by meridians. This projection is widely used for topographic maps from 1:25,000 scales to 1:250,000 scale, and it is the basis of the UTM coordinate system.

Lambert Conformal Conic

The Lambert Conformal Conic (LCC) projection, presented by Johann Heinrich Lambert in 1772, shows meridians as equally spaced straight lines converging at one of the poles. Angles between the meridians on the projection are smaller than the corresponding angles on the globe. Parallels are unequally spaced concentric circular arcs centred on the pole, and spacing of the parallels increases away from the pole. The pole nearest the standard parallel is a point and the other pole cannot be shown. The scale is true along the standard parallel or along two standard parallels and is constant along any given parallel. The LCC projection is extensively used for large-scale mapping of regions with an elongated axis in the East- West directions and in mid-latitude regions. It is standard in many countries for maps at 1:500,000 scale, as well as for aeronautical charts of a similar scale.

Stereographic

The Stereographic projection, developed by the 2nd century B.C., is a perspective azimuthal projection that preserves angles (i.e., is conformal). This projection is the only projection in which all circles from the globe are represented as circles in the plane of projection. The polar, Equatorial and oblique aspects result in different appearances of the graticule. The polar aspect is achieved by projecting from one pole to a plane tangent at the other pole. In this aspect, meridians are equally spaced straight lines intersecting at the pole with true angles between them. Parallels are unequally spaced circles centred on the pole represented as a point. Spacing of the parallels increases away from the pole. The Stereographic projection is used in the polar aspect for topographic maps of Polar Regions. The Universal Polar Stereographic (UPS) is the sister projection of the UTM for military mapping. This projection generally is chosen for regions that are roughly circular in shape. It is in use in oblique ellipsoidal form in a number of countries throughout the world, including Canada, Romania, Poland and the Netherlands. Different countries have different mathematical developments or versions of the Stereographic projection.

Preserving Areas

Lambert Cylindrical Equal Area

The Cylindrical Equal Area projection was first presented by Johann Heinrich Lambert in 1772. It became the basis for many other similar equal area projections including the Gall Orthographic, Behrmann, and Trystan-Edwards projections. Lambert's original projection uses a single line of constant scale along the Equator. Similar equal area projections are constructed using two parallels as the lines of constant scale. On the Lambert Cylindrical Equal Area projection, meridians are equally spaced straight parallel lines and the Equator is π times as long as the meridians. Lines of latitude are unequally spaced parallel lines furthest apart near the Equator and are perpendicular to the meridians. Changing the spacing of the parallels is the method used to preserve equal areas. Significant distance and angle distortion, however, results with the distortion greater in high latitudes near the poles. This projection is not often used directly for map construction, but it is a standard to describe map projection principles in textbooks and has also served as a prototype for other projections.

Mollweide

In 1805, Carl Brandan Mollweide developed a pseudo cylindrical equal area projection on which the central meridian is a straight line one-half as long as the Equator forming an elliptical area of projection for the entire globe. The meridians 90° East and West of the central meridian form a circle on the Mollweide projection. Other meridians are equally spaced semi ellipses intersecting at the poles and concave toward the central meridian. Parallels are unequally spaced straight lines and are perpendicular to the central meridian. The parallels are farthest apart near the Equator with spacing changing gradually.

Compromise Projections

Map projections that are neither conformal nor equal area are called compromise projections. They are almost unlimited in variety. Among them are many important and useful projections.

Orthographic

The Orthographic projection, developed by the 2nd century B.C., is a perspective azimuthal projection that is neither conformal nor equal area. It is used in polar, Equatorial and oblique aspects and results in a view of an entire hemisphere. The polar aspect of the projection has meridians that are straight lines and intersect the central pole with the angles between meridians being true. The pole is a point and the parallels are unequally spaced circles centred on the pole. The spacing parallels decrease away from the pole. Scale is true at the centre and along the circumference of any circle with its centre at the projection centre. The projection has a globe-like look, and is essentially a perspective projection of the globe onto a plane from an infinite distance (orthogonally). It is commonly used for pictorial views of the Earth as if seen from space.

Gnomonic

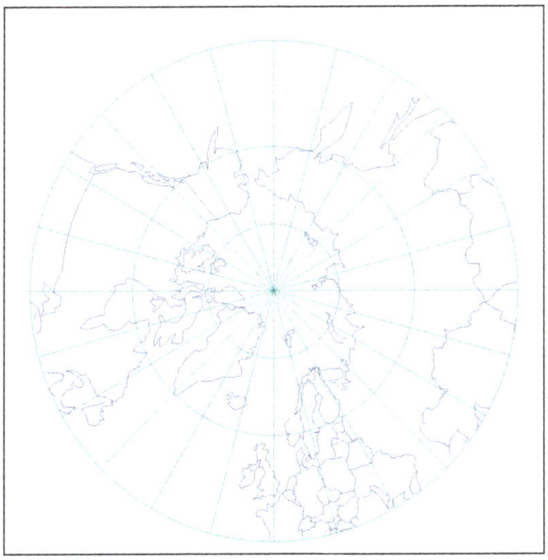

The Gnomonic projection, which maps great circles to straight lines.

The Gnomonic projection is neither conformal nor equal area. It is a perspective azimuthal projection with the point of projection at the centre of the Earth, which is the source of the name (i.e., the centre of the Earth where the mythical gnomes live). It was developed by the Greek Thales, possibly around 580 B.C. All great circles on the projection, including all meridians and the Equator, are shown as straight lines, a property unique to this projection.

The graticule appearance changes with the aspect, as with other azimuthal projections. Meridians are equally spaced straight lines intersecting at the pole with true angles between them in the polar aspect. Parallels are unequally spaced circles centred on the pole as a point, and the spacing of the parallels increases from the pole. The projection only can show less than a hemisphere. Scale increases rapidly with distance from the centre. Its usage results from the special feature of representing great circles as straight lines, and it thus assists navigators and aviators in determining the shortest courses.

Azimuthal Equidistant

In this polar aspect projection, meridians are equally spaced straight lines intersecting at the central pole. Angles between them are the true angles. Parallels are equally spaced circles, centred at the pole, which is a point. The entire Earth can be shown, but the opposite pole is a bounding circle having a radius twice that of the Equator. In its equatorial aspect, meridians are complex curves, equally spaced along the Equator and intersecting at each pole. Parallels are complex curves concave toward the nearest pole and equally spaced along the central meridian and the meridian 90° from the central meridian. The scale is true along any straight line radiating from the centre of projection. It increases in a direction perpendicular to the radius as the distance from the centre increases. Distortion is moderate for one hemisphere but becomes extreme for a map of the entire Earth. The distance between any two points on a straight line passing through the centre of projection is shown at true scale; this feature is especially useful if one point is the centre.

This projection is commonly used in the polar aspect for maps of Polar Regions, the Northern and Southern Hemispheres, and the "aviation-age" Earth. The oblique aspect is frequently used for world maps centred on important cities and occasionally for maps of continents. The Azimuthal Equidistant projection was recognized by the UN and used on the UN's flag.

The azimuthal equidistant projection for preserving distances on the UN's flag.

Winkel Tripel

The Winkel Tripel projection is neither conformal nor equal area. It was presented by Oswald Winkel of Germany in 1921. The projection was obtained by averaging coordinates of the Equidistant Cylindrical and Aitoff projections. Winkel applied the name "Tripel," normally meaning triple, because the Aitoff projection is an equatorial aspect of one hemisphere of the Azimuthal Equidistant projection, on which horizontal coordinates have been doubled and meridians have been given twice their original longitudes.

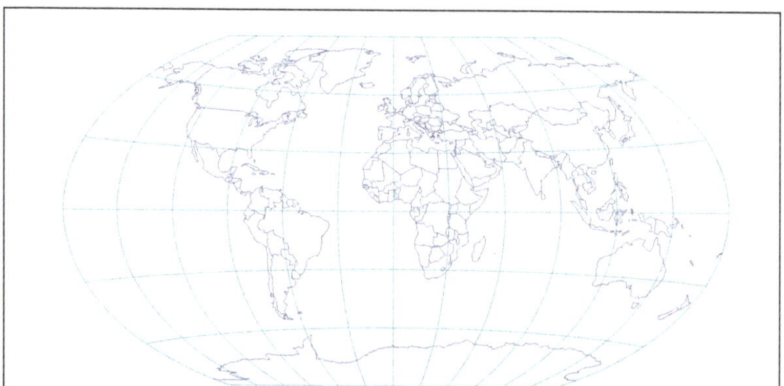

Winkel Tripel projection.

The central meridian is straight. Other meridians are curved, equally spaced along the Equator and concave toward the central meridian. The Equator and the poles are straight. Other parallels are curved, equally spaced along the central meridian and concave toward the nearest pole. Poles are straight lines about 0.4 as long as the Equator, depending on the latitude of the standard parallels. Scale is true along the central meridian and constant along the Equator. Distortion is moderate except near outer meridians in Polar Regions. The Winkel Tripel is used for whole-world maps.

Modern Approaches to Map Projections

Web Mercator

Many major online street mapping services (Bing Maps, Open Street Map, Google Maps, MapQuest, Yahoo Maps, and others) use a variant of the Mercator projection for their map images. Despite its obvious scale variation at small scales, the projection is well suited as an interactive world map that can be zoomed into seamlessly to large-scale (local) maps, where there is relatively little distortion due to the variant projection's near-conformity.

The scale factor at a point on a conformal map projection (such as the spherical Mercator or the ellipsoidal Mercator) is uniform in all directions. This is not true on a Web Mercator. Let us denote with m the scale factor in the N/S meridian direction and with n the scale factor in the E/W parallel direction. Then m = n because the scale factor at a point is the same in all directions on the spherical Mercator projection. In other words, the spherical Mercator is conformal.

The equations for the ellipsoidal Mercator are a little more complicated, especially in Northing. The parameters a (semi-major axis) and e (eccentricity) are given for the selected ellipsoid. Again m = n because the scale factor at a point is the same in all directions on the ellipsoidal Mercator projection. In other words, the ellipsoidal Mercator is conformal.

Web Mercator is the mapping of WGS84 datum (i.e., ellipsoidal) latitude/longitude into Easting/Northing using spherical Mercator equations (where R = a). This projection was popularized by Google in Google Maps (not Google Earth). The reference ellipsoid is always WGS84, and the spherical radius R is equal to the semi major axis of the WGS84 ellipsoid a. That's "Web Mercator."

The scale factor at a point is now different for every direction. It is a function of the radii of curvature in the meridian and the prime vertical and the direction alpha. For the Web Mercator, m and n are not equal. Thus, the Web Mercator is not a conformal projection. If somebody uses the Web Mercator for printing out directions to a new restaurant across town or for visualization on his/her computer screen or for other purposes on the web, there will be no problem. But the Web Mercator is a projection that has jumped from one domain of use (the web) to another domain of use (GIS) where it is leading another life. Witnesses are the EPSG, Esri and FME codes for the Web Mercator. Surveyors and GIS professionals need to know that the Web Mercator is not conformal. If distance computations on the Web Mercator are done simply (as they can be done on a conformal projection), they will be wrong. If done correctly, they will be laborious.

For an area the size of the NW quadri-sphere (North America), the differences appear light. It turns out that the Eastings are identical. The differences are in the Northings. There is no Northing difference at the Equator, but by 70 degrees North, the difference is 40 km. This NS stretching in the Web Mercator is the reason for its non-conformity. Mercator projections are useful for navigation because rhumb lines are straight. These are lines of constant true heading that navigators used to sail before GPS. So, we have to have in mind that straight lines on a Web Mercator are not rhumb lines. To summarize about the Web Mercator:

- The Web Mercator is cylindrical.
- Its meridians are equally spaced straight lines.
- Its parallels are unequally spaced straight lines but in a different way than a conformal Mercator.
- Its loxodromes (rhumb lines) are not straight lines.
- It is not perspective.
- Its poles are at infinity.
- It was not presented by Mercator, but by Google recently.
- It is not conformal.

Map Projection Transitions

Map Projection Transitions is an example of multiple applications offered by Jason Davies. The map is not static, but animated. The South Pole moves toward the bottom and Earth rotates around its poles. The animation lasts five seconds, after which the projection changes and movement continues for five seconds, after which the projection changes again. Names of projections appear in a separate window. There are a total of 56 projections. The South Pole eventually becomes invisible and the North Pole appears at the top. Various parts of Earth appear in the centre of the map by rotating around the poles.

By clicking Pause, animation stops and it is possible to select another projection. By left-clicking, it is possible to move the picture around and select projection aspect— normal, transversal or any of numerous oblique projections. Differences between two projections can be seen clearly in such a way. For example, one is able to select the Ginzburg VI projection and its normal aspect by moving the mouse. If one wants to see how that projection's graticule is different from the similar Winkel Tripel projection, it can be done by clicking on the Winkel Tripel projection on the drop-down menu. The picture on the screen is going to change to the Winkel projection and differences are going to be clear.

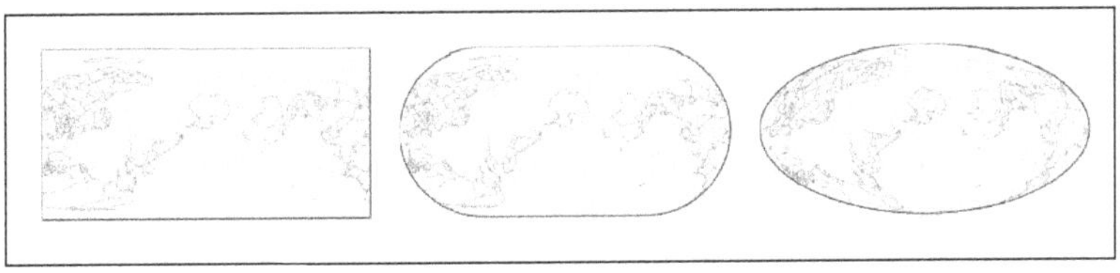

From the Map Projection Transitions application.

If one clicks on Maps, there is a series of new interesting applications about interrupted maps, butterfly-shaped maps, retro-azimuthal projections and other projections. It is possible to use the mouse to move pictures in many of those applications. For example, by selecting the interrupted sinusoidal projection, a world map in three segments is going to appear. The mouse can be used to move parts of Earth from one segment to another, and the slider at the bottom of the screen can be used to change the number of segmentsfrom an uninterrupted world map to a representation in 24 segments.

A similar option is available for the Berghaus star projection. The application Azimuth and Distance from London enables using the mouse to obtain distances and azimuths from London to any point on Earth in world maps in oblique equidistant cylindrical and oblique equidistant azimuthal projection.

Map Designing

The term design normally refers to the planning of a map particularly to the choice and arrangement of its graphic elements. For GIS, graphic output must show the features appropriately symbolized or labeled and the relationships are discernible. The map design process includes making decisions about the selection of geographic features and attributes to be represented on the map, selection of an appropriate map projection given the purpose of the map and the geographic region of interest, choice of an appropriate scale, selection of appropriate symbolization, and design of the overall layout of the map.

The quality of map to a large extent depends on the quality of hard copy maps. Therefore, design of graphic output is critical if accurate information is to be conveyed effectively to the user. The map design process is especially important because maps are a synoptic form of communication. The aim of cartographic design is to present the geographical data in such a fashion that the map as a whole appears as an integrated unit and each item is clear and legible. Graphic output from GIS is often poorly designed. For example, there rarely exists any appropriate scaling of colours or patterns. The principles of cartographic design are not considered at all. The map is what distinguishes GIS as a different approach to the management of information, so extra care should be taken to improve the final maps that a GIS generates. The issues of graphic design in GIS has become more important because of the fact that screen display introduces a new set of issues as it has greater capabilities compared with paper maps. The concepts of designing are about the following three aspects which are as follows:

- The purpose of design is to capture the attention of the user.
- The Principles of Cartographic Design are Timeless, the Results are not.
- The Rules of Cartographic Design can be taught and learnt; principles and concepts have to be acquired.

There are five basic principles of map designing. These are:

- Concept: Concept here means that the developer should have a clear idea about the purpose of the map and the nature of its user. While designing, the entire objective be incorporated and it must be complete in itself. Design comes in two stages, concept and parameters, and detail in execution.
- Hierarchy: Important things must look important, and the most important thing should look the most important. Lesser things have their place and should serve to complement the more important aspects.
- Simplicity: The beauty of designing that is simple and straight forward. Maps are developed for others so the important aspect of it is not what you put in that makes a great map but what is inferred from it. This is the designer's skill. Content may determine scale or scale may determine content, and each determines the level of generalization.
- Maximum Information: How much information can be gained from this map, at a glance? Functionality not utility is the key. Design makes utility functional. All designs are a compromise. The spark which makes a map special often only comes when the map is complete.
- Understanding: Good designers use Cartographic fictions, Cartographic impressions, and Cartographic illusions to make a map. The image is the message. Good design is a result of the tension between the environment and the designer. Design uses aesthetics but the principles of aesthetics are not those of design.

Elements of Map Design

The features, symbols, and text, their relative placement and, the process of choosing the information to be conveyed and the audience for which it is intended, balance, contrast etc. are the key elements of map designing. The proper design for other types of geographic information (charts, graphs, etc.) can be considered as a sub-set of proper map design. Some elements are found on almost all maps no matter what the theme, others depend heavily on the context in which the map will be read. The best GIS database may be a failure, if the information it contains is not displayed effectively.

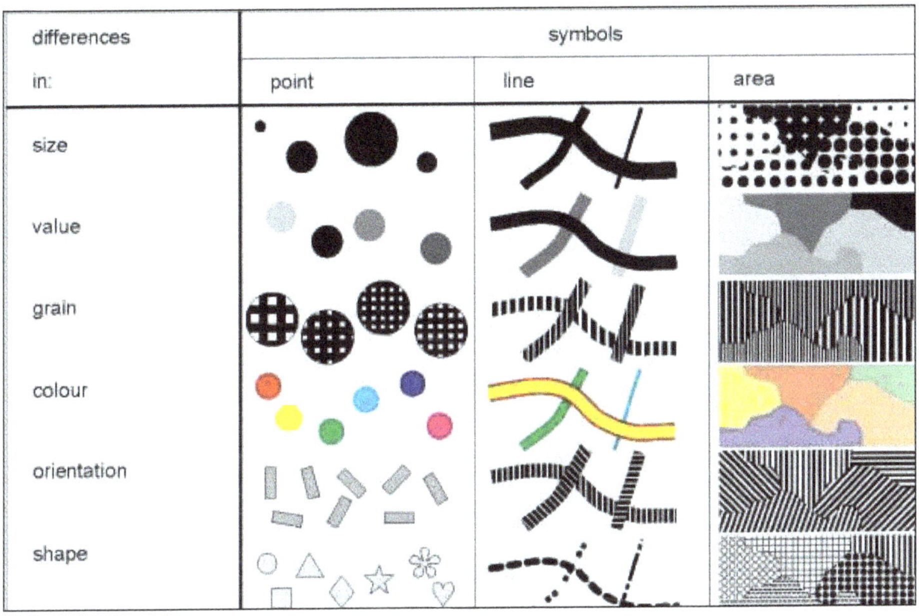

Contrast

No element of cartographic technique is as important as contrast. Sharpness of a map is determined by the contrast. Contrast may be achieved by varying the appearance of lines, shapes, size, patterns, colour, value (intensity) and direction. A map has number of linear features such as coastlines, rivers, railway lines, roads, and various political boundaries. In order to distinguish them; it is necessary to vary their character, size, shape and pattern. Visual contrast can be based on any of the graphic variables. Though visual contrast is necessary and can enhance the interpretability of the map, but too much of it can detract from the clarity and order of the map.

Pattern

Patterns are composed of dots, lines or combination. It has the following characteristics:

- Texture: It relates to the spacing of the marks and usually specified by the number of lines/dots per inch. The greater the number, the finer is the texture.
- Arrangement: It refers to the relationship of the positioning of the marks, such as parallel lines, crosshatched, arrangement of dots. It must be noted that while placing dots, don't follow any defined pattern.
- Orientation: It refers to positioning of the arrangement relative to the viewer as vertical, horizontal or diagonal.
- Value refers to the relative darkness.

Use of Colour

Color		Byte binary (red, green blue)
White		255, 255, 255
Gray		127, 127, 127
Black		0, 0, 0
Red		255, 0, 0,
Green		0, 255, 0
Blue		0, 0, 255
Yellow		255, 255, 0
Cyan		0, 255, 255
Magenta		255, 0, 255
Orange		255, 153, 0
Pink		255, 170, 170
Purple		170, 0, 170
Teal		0, 170, 153
Brown		153, 102, 51
Tan		255, 204, 102

Munsell colour scheme.

The use of colour on maps serves more purposes than to just construct a legend. Colour conveys visual meaning. It directs attention and it can emphasize or obscure features. Thus, colour selection is an important element of map design. The human eye can distinguish tens of 1000s of different colors, but only 20-30 gray tones. The eye is also able to identify patterns and boundaries (edges) and detect movement.

On a static map, however, the eye cannot reliably match more than 16-20 simple colours from the legend of the map. This limited ability can be expanded by the addition of pattern elements (symbols) to the basic colours. Still, there is a practical limit to how many colours should be displayed on a map. Maps are easiest to interpret when the number of colours does not exceed 10 to 12. Too many colours (>20) create a map that is visually cluttered.

The nature of colour and human perception of colour are one of the most important elements in cartography. Blue, green and red are the primary colours, also known as additive colours because they can be added together to create all visible colours of the spectrum. Colours may also be created by subtracting an additive colour from white. For example, white-red = cayon = (blue + green).

Colour printing (newspapers, books) is based on subtractive colors (plus black), whereas color video display (television, monitors) is based on additive colours. In printing a colour document from a computer display, the printer must translate additive colours into corresponding combinations of subtractive colours. The Munsell colour code is the internationally accepted standard for describing colour quantitatively.

Colour has three major aspects namely, hue value, percent reflectance, lightness or brightness and Chroma also termed as saturation or intensity. These three have been employed in the Munsell system. When we speak red, green or blue, in fact we are describing the hue. Value of the colour is sensitivity of the colour. The tones of the scale value range from black to white and referred as gray scale. Chroma refers to the reactions of the eye to the colour. It also refers to saturation, which is the amount of hue in a colour. The colours on monitor are additive colours. When all the three basic colours are added, this will lead to a white sensation. The standard VGA colour monitor can display 16 colours from a total colour selection of 262,144 (64^3). Super VGA monitors can display up to 256 simultaneous colours from a total of several million possibilities. More advanced monitors can display still more colours.

Colour may carry standardized meanings in certain specific applications. For example, blue for water bodies, green for vegetation, brown for desert/mountains (contours), white for ice or snow, etc. A standard colour scheme is often used on geological maps: green for Cretaceous, yellow for quaternary, blue-green for areas below sea level, green and yellow for lowland and hills, brown for mountains, red for forest areas over 5000 meters etc. The effect was meant to be that areas appeared to be nearest to the viewer were in fact farthest from the sea level.

The sequence of colours in a legend serves an important function in showing the visual relationship of map features. A smooth sequence of colours best shows data that are continuously variable. It is recommended that instead of varying colours we should vary the tones. Dissimilar colours create visual barriers between features that are not related to each other.

Balance

Balance here means, distribution of the elements as evenly as possible within the map frame to avoid unnecessary crowding and conversely, large blank areas. In other words, the relationships between the visual components should appear logical and the map should be legible. The most important elements in a given design should be featured in more prominent positions and occupy a larger area within the frame. It is advised that the most important information should be placed near the top or to the left of the map. Less important and ancillary map elements can be positioned toward the bottom and right. In this way, the importance of the various map elements can be matched to the visual hierarchy of the map itself. Visual balance is affected by:

- The "weight" of the symbols.
- The visual hierarchy of the symbols and elements.

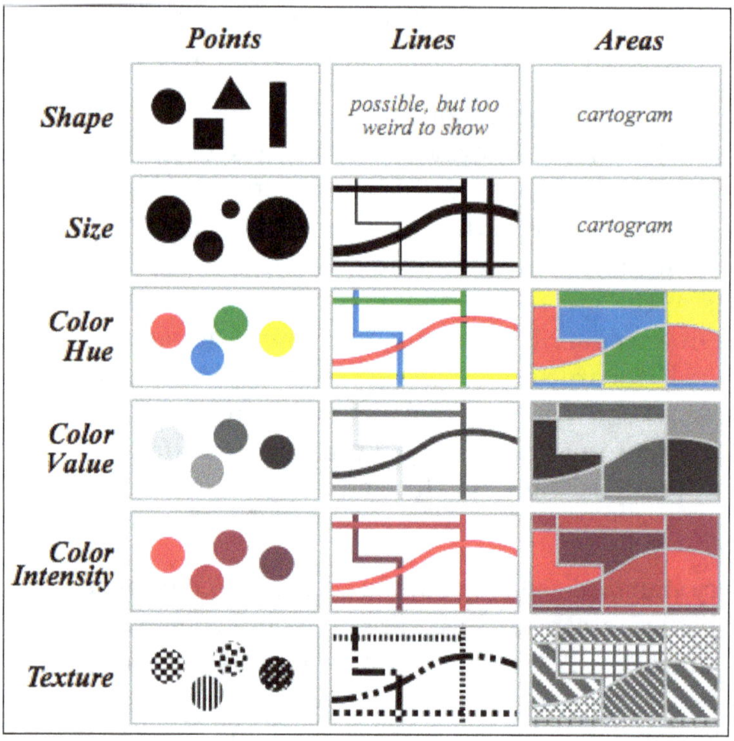

Title

The title should be written in a larger, bold font than any other supporting map text or annotation for greater emphasis. The title has to be placed at the top of the map. It is often useful to include the location of the map elements in the title. Use of abbreviations or unspecified acronyms in the title should be avoided.

Legend

The map legend explains the meaning of the point, line, and area symbols used to represent features on the map. Every symbol that appears on the map should necessarily appear in the legend, and it must appear exactly as it does in the body of the map. Legends are generally used for depicting qualitative data whereas for quantitative data the term index is preferred.

Visual Hierarchy

Perhaps the most important design decision is to ensure that the relative importance of information shown on the map is reflected in the choice of symbolization. Unnecessary information should be eliminated and relevant information be symbolized in a way that makes more important information visually more prominent. This requires careful consideration of the map purpose as maps with different purposes may display the same geographic features but with different emphasis.

The visual hierarchy of a map can be enhanced by manipulating figure-ground relationships. The human eye-brain system automatically separates visual images into background and foreground. In the context of maps, the background is typically base map information, which shows the spatial context for the thematic content of the map. Figure-ground relationships can be manipulated to distinguish land from water, or the thematic content of the map from the base map.

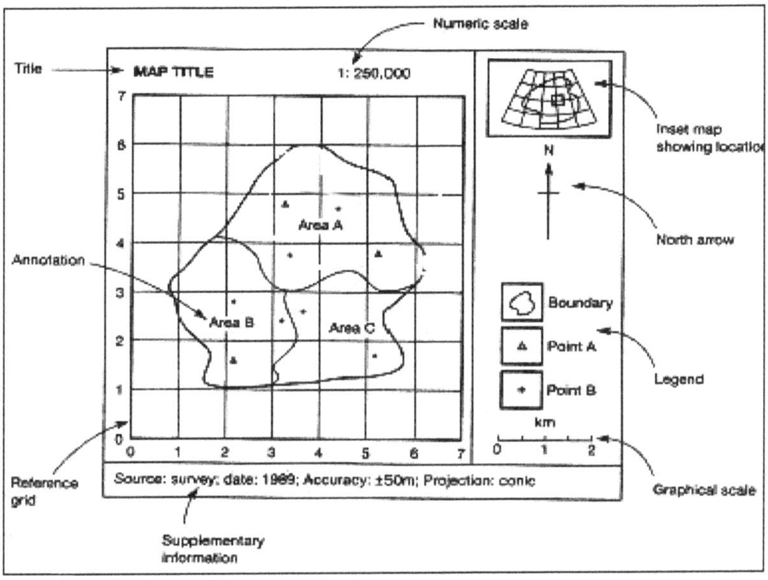

Map Elements.

Location

It is imperative to include some description of the location of the features on the map. The location of the map features may be given in the form of geographical or some other well accepted coordinate system. Besides, the inset map may also be given along with the map for displaying the location of the map within a larger area on small scale.

Every map is a collection of information. All sources of map information must be referenced. The source should include the date, title, author, publisher, and any other known and other pertinent information.

Projection

Almost all printed output of geographic data is projected onto a flat surface. Since this transformation results in some kind of distortion, it is mandatory that the projection system be mentioned so as the type and distortion can be assessed. It should include; name of the projection, central meridian, zone, the datum and spheroid. In the case of a large-scale map, which covers a small area, the map projection may be omitted from the map.

Text Characteristics

There are several text characteristics that can be manipulated by the map designer. Text style is commonly used to distinguish between different types of features. For example, upright characters are generally used to represent land features, while italics are often used to label water bodies. The form of text characters can be used to indicate the relative importance of features. Form refers to the text font. Although there are many fonts available, they can be grouped into two broad classes: serif or sans serif. Serifs are semi-structural details on the ends of some of the strokes that make up letters and symbols (also known as Roman).

Serif letter.

Sans serif fonts are the simplest. Arial is the typical example of Sans serif font. These fonts are often used on maps to label relatively minor features. A typeface without serifs is called sans-serif, from the French sans, meaning "without" (also known as Gothic). In traditional print, sans-serif fonts are more typically used for headlines than for body text. Sans-serif fonts have become the de facto standard for body text on-screen, especially online.

Sans-serif letter.

Serif fonts such as Times New Roman may be used to identify features of greater significance. For example, cities might be labeled using a sans serif font but capital cities might be labeled with a serif font. The size of the letters can also be used to indicate importance. Size refers to both the height of the letters (point size) and their weight (normal, bold). The spacing of lettering can be used to indicate feature extent. Point and line features are generally labeled using closely spaced lettering, while areal features are labeled with widely spaced lettering to show the extent and orientation of the feature. Placement and orientation of the text is used to indicate the location and extent of features.

Placement of the Text

Placement of lettering on maps is more important on reference maps that typically label many geographic features than on thematic maps. On thematic maps, text is often limited to the title, legend and other peripheral information. Labeling of features on the map involves trade-offs between locational precision and legibility. Placing of label points on maps is an important aspect of map design as they provide the greatest flexibility to attach descriptions to point, line and area features. If properly placed, it identifies the phenomenon to which it refers. In contrast, poorly placed label may carry some other meaning. For making a map user friendly and meaningful, standardized format for label placement has been developed. Some general rules for text placement are:

- Names should be legible and easily associated with the features they describe.
- Do not break names.
- Do not extend names along rivers or other line features.
- If possible, Place River names in downstream direction.
- Closely associate name with feature.
- Do not bend names excessively.
- Place so as to show the extent of the feature.
- Reflect the hierarchy of features by the use of different font sizes.
- Not be densely clustered nor evenly dispersed.

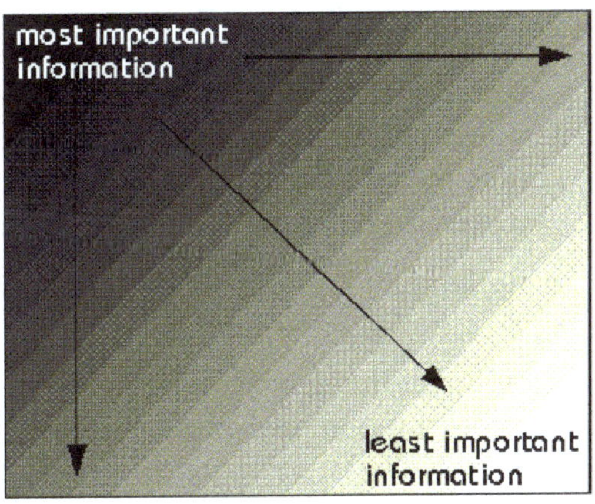

- Names should be either entirely on the land or on the water.
- Lettering should generally be oriented to match the map orientation. That is, it must be parallel to the upper and lower edge of the map for the large scale and parallel with the parallels for the small-scale maps.
- In case of conflict between lettering and data, the data should be interrupted not the names.
- Care must be taken that there is no over posting (overlap). Though labels must be positioned to avoid over posting, but without destroying the eye's ability to associate labels with appropriate features.

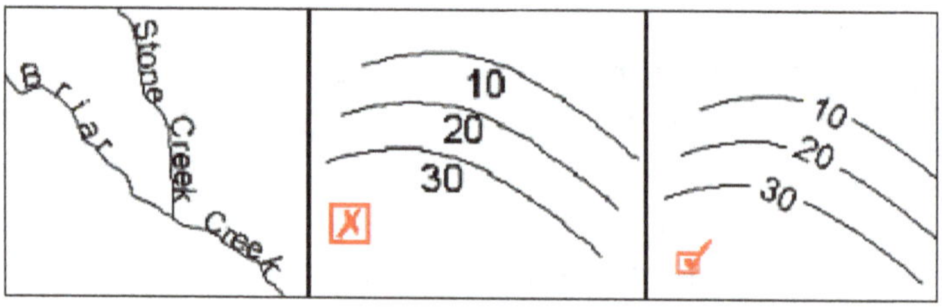

Line labelling.

Point Labeling

Optimum position for a label is above and to the right of the point. Below and to the right is less acceptable. The least acceptable position is to the left and below of the point. Label can be turned (non-horizontal) if necessary, but only by a small amount.

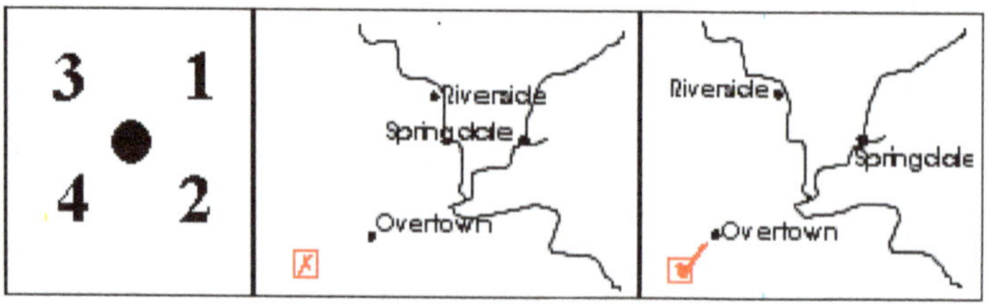

Point Labeling.

Polygon Labeling

Labeling of polygons within automated mapping is difficult and challenging programming problem. The label should be central to the feature, may be reoriented or curved to fit the feature. In some cases the label may be connected with the feature by an arrow.

Diagram Labeling

Labeling of diagram is generally a matter of trial and error. Search for feasible positions for a rectangle wholly enclosed within the polygon. The ratio of width to height should be as high as possible. If the nature of the data suggests the shape of the graphic, follow that suggestion - otherwise, move toward horizontal graphics about 50% wider than tall. While mapping, always follow the traditional cartographic symbols and colours. For example, water in blue colour, forest as green, contours in brown, grid lines in purple, cropped areas as yellow etc. By straying from these traditional methods of symbolization the map may be misinterpreted.

Map Algebra

Map algebra is an informal and commonly used scheme for manipulating continuously sampled (i.e. raster) variables defined over a common area. It is also a term used to describe calculations within and between GIS data layers, according to some mathematical expression, to produce a new layer; it was first described and developed by Tomlin. Map algebra can also be used to manipulate vector map layers, sometimes resulting in the production of a raster output. Although no new capabilities are brought to GIS, map algebra provides an elegant way to describe operations on GIS datasets. It can be thought of simply as algebra applied to spatial data which, in the case of raster data, are facilitated by the fact that a raster is a geo referenced numerical array.

Map Algebra models the surface of the earth as a multitude of independent, coincident layers or themes. The layers interact according to mathematical models and are typically based on real world observations. Planners develop layers on development and population. Social scientists develop layers on demographics, ethnicity, and economic factors. Applying Map Algebra model to input layers produces a new layer, which may be a physical map sheet, a vision perceived through a stack of mylars on a light table, or an electronic dataset displayed on a computer screen. Regardless of mechanism, the result allows its users to explain complex phenomena, predict trends, or make adjustments to the model.

However, it is the mechanism which bounds usability of Map Algebra. How easy it is for scientists to perform simple tasks? Can complex models be developed and tested? Historically layers were plotted on individual transparent maps which, when superimposed and registered provide a visually integrated view of the data. The manual process of map overlay is slow and tedious.

Types of Map Algebra

The most common type of map algebra is a cell-by-cell function. This type has rasters directly stacked on top of one another. Then, the function applies to cells aligned with each other. But you can also change the configuration so it's not only cell-by-cell. Here are the types of map algebra operations that you can use: Local, Zonal, Focal and Global.

Local Operations

The simplest approach is map algebra on a cell-by-cell basis. For example, you have 2 rasters stacked on top of each other. Then, you add each cell one-by-one. If raster cells represent temperature, you can subtract them both at different time periods. By using a local operation, you can find the difference in temperature for each cell.

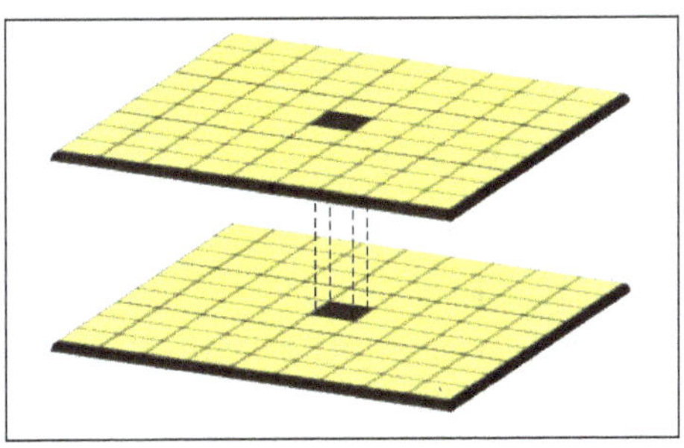

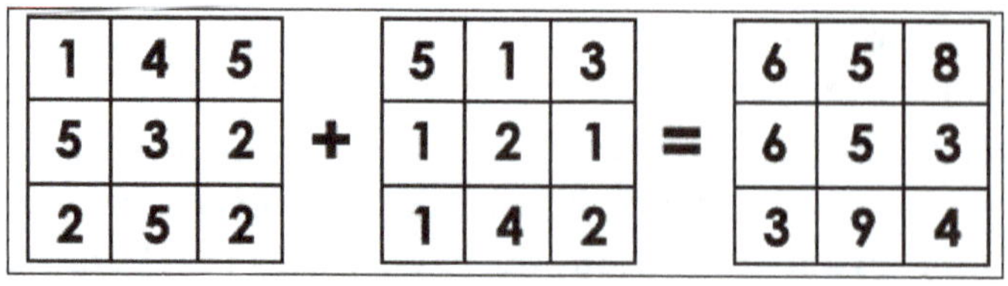

Global Operations

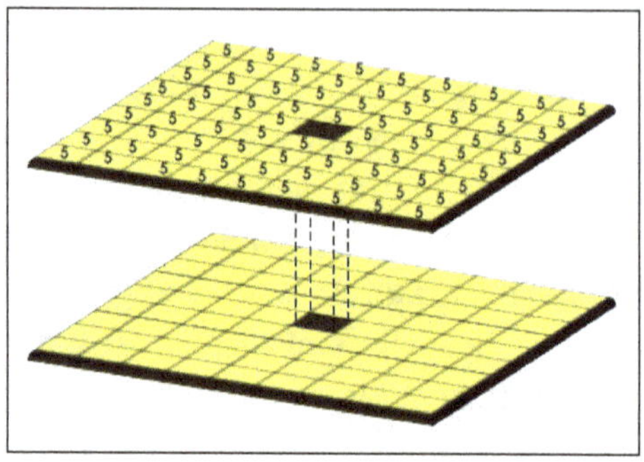

Global operations apply a bulk change to all cells in a raster. If you want to add a value of 1 to all grid cells, this is a global operation. For example, Euclidean distance is an example of a global operation. By calculating the closest distance away from a source, it applies the function globally in a raster.

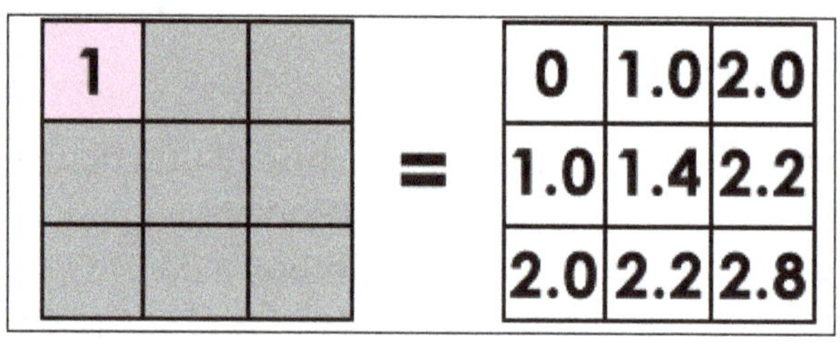

Focal Operations

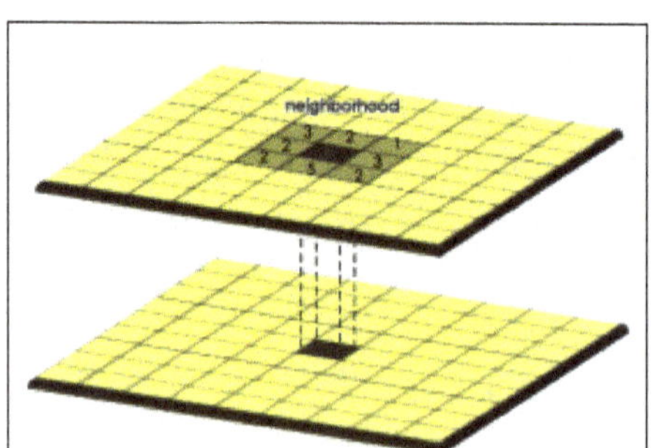

Focal operations are spatial functions that compute an output value of each cell using neighborhood values. For example, convolution, kernel, and moving windows are focal operations.

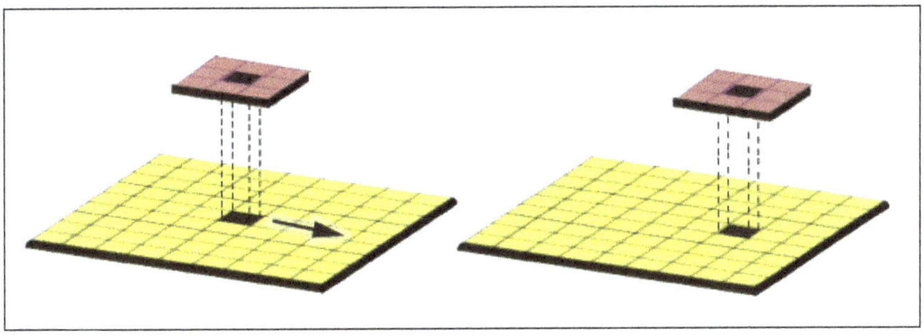

Moving window: A moving window is a rectangular arrangement of cells that shifts in position. By applying an operation to each cell from a moving window, it commonly smooths values in a raster.

Zonal Operation

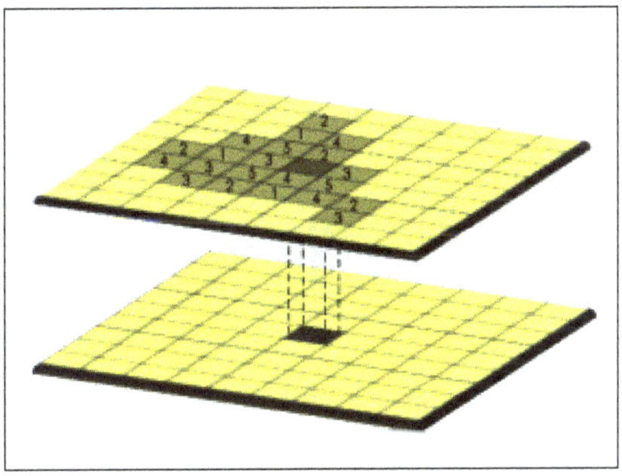

Zonal operations apply a math function to a group of cells within a specified zone. For example, a zone could be a vector or raster format. If you want to find out how much rain is in a watershed, you can use a zonal operation. By using a raster with precipitation, you can set your zone as a watershed with a sum function.

Mathematical Functions

Here are examples of operations that you can use for map algebra:

- Arithmetic operations (addition, subtraction, multiplication, division).
- Statistical operations (minimum, maximum, average, median).
- Relational operations (greater than, smaller than, equal to).
- Trigonometric operations (sine, cosine, tangent, arcsine).
- Exponential and logarithmic operations (exponent, logarithm).

Map Algebra Example

If you were to take a snapshot of land surface temperature in April 2015, it would look like this:

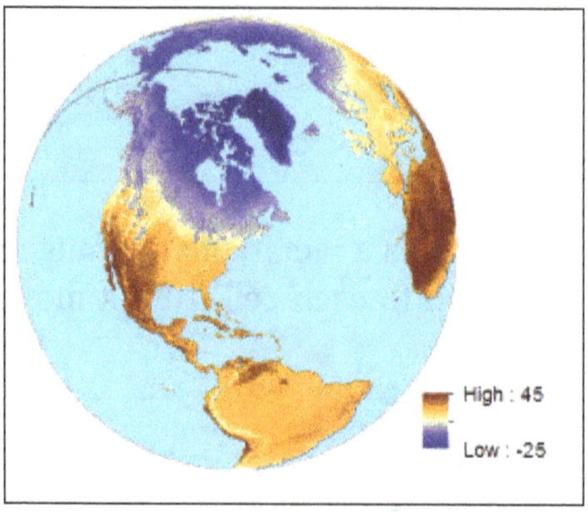

Then, if you took April 2014 land surface temperature, here's what it looks like:

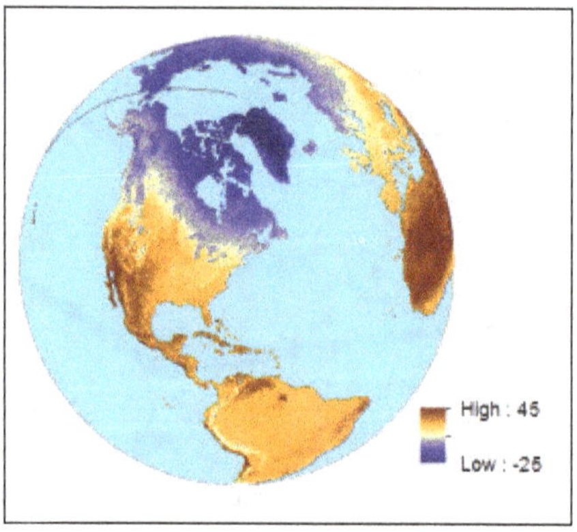

When you subtract April 2014 from April 2015 for land surface temperature, it shows the difference between temperatures for two given periods.

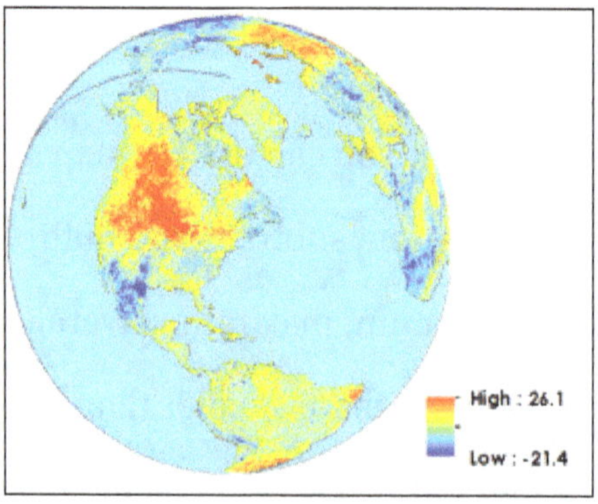

This is an example of a local operator because it subtracts each cell from one year to another. It's also an example of an arithmetic operator because it uses subtraction.

Cells overlap with a range of -25° to +45 °C. Then, we subtracted 2014 from 2015 land surface values for each cell. As a result, the output map has a range from -40° to +40 °C. What's special about this is that you can pinpoint cells with the greatest difference. If values are positive, this means that 2015 land surface temperatures were hotter. But if values are negative, this means that 2015 land surface temperatures were colder. What happens when you have null values? Null isn't a numerical value. If there are null values in any of the raster cells, they will remain as null in the output raster.

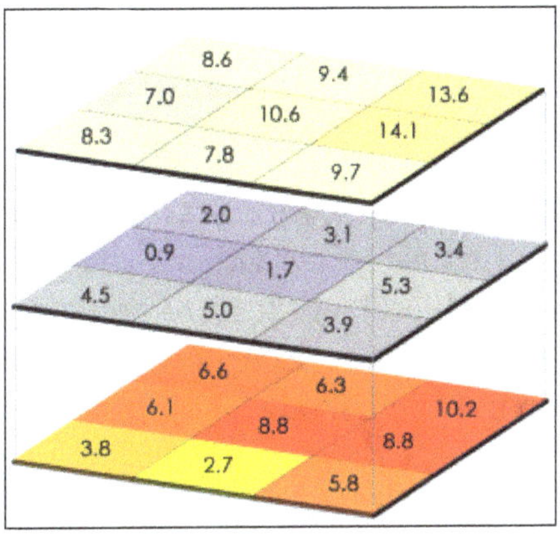

Relational Operators

Relational operators are logical functions that test a relationship. For example, here are operators that test logic. Examples of relational operators are:

- Equal to
- Not equal to
- Less than
- Greater than

If it's true, it returns a value of 1. But if it's false, then it returns a value of 0.

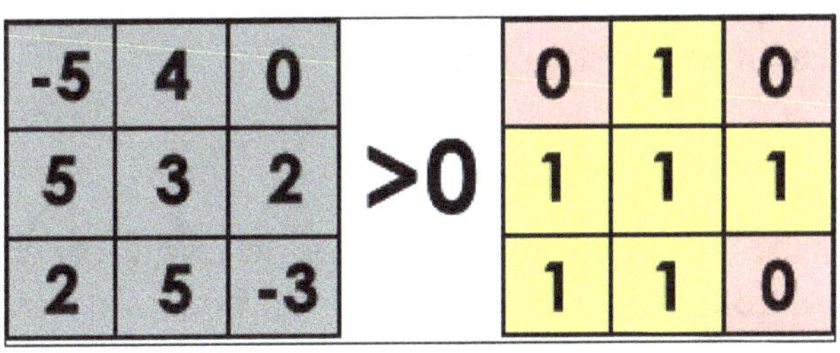

Statistical Operators

Statistical operators calculate statistics for each cell by using statistical functions. For example, minimum, maximum, average, or median are statistical operators. You can really crunch numbers using tools like zonal statistics. This tool uses a statistical operator to evaluate numbers for all cell values within a zone. For example, zonal mean, median, and minimum are tools that use statistical operators.

Data Types

Here we focus on map algebra operations available for gridded datasets, such as those implemented in ESRI's Spatial Analyst extension for ArcView. Similar operations are available in other grid-based GISes, such as GRASS (Geographic Resources Analysis Support System). The data associated with any grid cell can be of any type whatsoever. It is conceptually useful to divide data types into several classes, however. These include:

- Categorical data: These are non-numerical data. Grids that classify land use or land cover exemplify this category. Other examples are proximity grids (values identify the nearest object) and feature grids (only two values are possible: one value for cells where features occur, another value--typically zero or NoData--where features do not occur).
- Integral data: These data may be relative ranks or preferences or they may be counts of occurrences or observations, for example. Thus, what they measure is inherently integral.
- Floating - point ("real" data): These typically represent a real surface, such as elevation, or the values of a scalar function (a "conceptual surface," if you will). Examples of such functions would be temperature, slope, amount of sunlight received per year, distance to the nearest feature, population density.
- Vector data: These are ordered tuples of real values that represent fields of directions. For example, hydraulic gradients (for two-dimensional groundwater models), wind velocities (again for two-dimensional models), and ocean currents are two-dimensional vector fields. Vector data may have more than two dimensions, even though they are defined over a strictly two-dimensional domain. For example, models using astronomical data, such as climate models, may make use of information about the three-dimensional location (on the earth's surface) of each grid point.

Working with Null Data

An essential part of map algebra or spatial analysis is the coding of data in such a way as to eliminate certain areas from further contribution to the analysis. For instance, if the existence of low-grade land is a prerequisite for a site selection procedure, we then need to produce a layer in which areas of low-grade land are coded distinctively so that all other areas can be removed. One possibility is to set the areas of low-grade land to a value of 1 and the remaining areas to 0. Any processes involving multiplication, division or geometric means that encounter the zero value will then also return a zero value and that location (pixel) will be removed from the analysis. The opposite is true if processing involves addition, subtraction or arithmetic mean calculations, since the zero value will survive through to the end of the process. The second possibility is to use a null or No Data value instead of a zero.

The null is a special value which indicates that there is no digital numerical value. In general, unlike zero, any expression will produce a null value if any of the corresponding input pixels have null values. Many functions and expressions simply ignore null values, however, and in some circumstances this may be useful, but it also means that a special kind of function must be used if we need to test for the presence of (or to assign) null values in a dataset. For instance, within ESRI's ArcGIS, the function ISNULL is used to test for the existence of null values and will produce a value of 1 if null, or 0 if not. Using ER Mapper's formula editor, null values can easily be assigned, set to other values, made visible or hidden. Situations where the presence of nulls is disadvantageous include instances where there are unknown gaps in the dataset, perhaps produced by measurement error or failure. Within map algebra, however, the null value can be used to great advantage since it enables the selective removal or retention of values and locations during analysis.

Table: Operations categorized according to their spatial or non-spatial nature.

Output	Spatial attributes involved?	
	Yes	No (not necessarily)
Map or image	Neighbourhood processing (filtering), zonal and focal operations, mathematical morphology.	Reclassification, rescaling (unary operations), overlay (binary operations), thresholding and density slicing.
Tabular	Spatial autocorrelation and variogram.	Various tabular statistics (aggregation, variety) and tabular modeling (calculation of new fields from existing ones), scattergraphs.

Logical and Conditional Processing

These two processes are quite similar and they provide a means of controlling what happens during some function. They allow us to evaluate some criterion and to specify what happens next if the criterion is satisfied or not. Logical processing describes the tracking of true and false values through a procedure. Normally, in map algebra, a non-zero value is always considered to be a logical true, and zero, a logical false. Some operators and functions may return either logical true values (1) or logical false values (0), for example relational and Boolean operators. The return of a true or false value acts as a switch for one or other consequence within the procedure. Conditional processing allows that a particular action can be specified, according to the satisfaction of various conditions; if the conditions are evaluated as true then one action is taken, and an alternative action is taken when the conditions are evaluated as false. The conventional if–then–else statement is a simple example of a conditional statement:

if i < 16 then 1 else null where i = input pixel dn

Conditional processing is especially useful for creating analysis 'masks'. In Figure below each input pixel value is tested for the condition of having a slope equal to or less than 15°. If the value tests true (slope angle is 15° or less), a value of 1 is assigned to the output pixel. If it tests false (exceed 15°), a null value is assigned to the output pixel. The output could then be used as a mask to exclude areas of steeper slopes and allow through all areas of gentle slopes, such as might be required in fulfilling the prescriptive criteria for a site selection exercise.

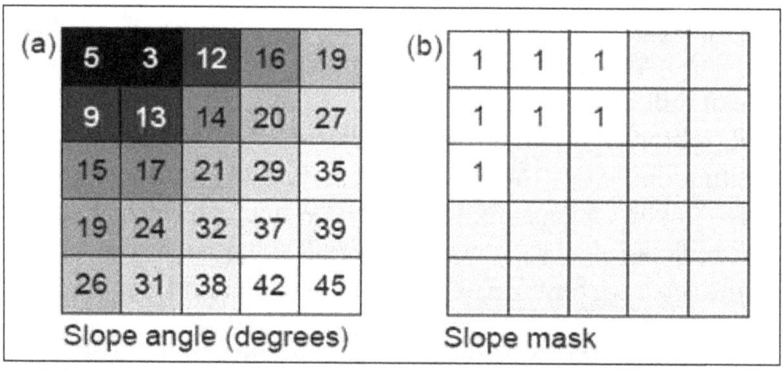

Logical test of slope angle data, for the condition of being no greater in value than 15°: (a) slope angle raster and (b) slope mask (pale grey blank cells indicate null values).

Other Types of Operator

Expressions can be evaluated using arithmetic operators (addition, subtraction, logarithmic, trigonometric) and performed on spatially coincident pixel DN values within two or more input layers. Generally speaking, the order in which the input layers are listed denotes the precedence with which they are processed; the input or operator listed first is given top priority and is performed first, with decreasing priority from left to right.

A relational operator enables the construction of logical functions and tests by comparing two numbers and returning a true value (1) if the values are equal or false (0) if not. For example, this operator can be used to find locations within a single input layer with DN values representing a particular class of interest. These are particularly useful with discrete or categorical data.

A Boolean operator, for example AND, OR or NOT, also enables sequential logical functions and tests to be performed. Like relational operators, Boolean operators also return true (1) and false (0) values. They are performed on two or more input layers to select or remove values and locations from the analysis. For example, to satisfy criteria within a slope stability model, Boolean operators could be used to identify all locations where values in one input representing slope are greater than 40° and where values in an elevation model layer are greater than 2000m.

Logical operators involve the logical comparison of the two inputs and assign a value according to the type of operator. For instance, for two inputs (A and B) A DIFF B assigns the value from A to the output pixel if the values are different or a zero if they are the same. An expression A OVER B assigns the value from A if a non-zero value exists; if not then the value from B is assigned to the output pixel. A combinatorial operator finds all the unique combinations of values among the attributes of multiple input rasters and assigns a unique value to each combination in the output layer. The output attribute will contain fields and attributes from all the input layers. All these operators can be used, with care, alone or sequentially, to remove, test, process, retain or remove values (and locations) selectively from datasets alone or from within a spatial analysis procedure.

Table: Description of Common arithmetic, relational, Boolean, power, logical and combinatorial operator.

Arithmetic	Relational (return true/false)	Boolean (return true/false)
+, Addition	= =. EQ Equal	^,Not Logical complement & AND Logical AND l, OR Logical OR !, XOR Logical XOR
-. Subtraction	^=,<>, NE Not equal	
*, Multiplication	<=.LT Less than/equal to	
/, Division	<=LE Less than/equal to	
MOD, Modulus	>,GT Greater than >=, GE greater than/ equal to	
Power	Logical	Combinational
Sqrt, Square root	DIFF, Logical difference	CAND, Combinational AND
Sqr, Square	IN{list}, Contained in list	COR, Combinational OR
Pow, Raised to a power	OVER, Replace	CXOR, Combinational XOR

Use of Boolean rules and set theory within map algebra; here the circles represent the feature classes A, B and C, illustrating how simple Boolean rules can be applied to geographic datasets, and especially rasters to extract or retain values, to satisfy a series of criteria: (a) A AND B (intersection or minimum); (b) A NOT B; (c) (A AND C) OR B; (d) A OR B (union or maximum); (e) A XOR B; and (f) A AND (B OR C).

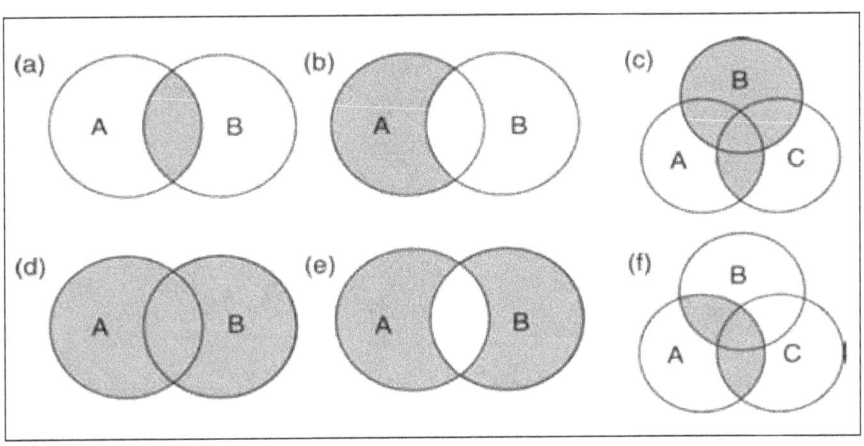

Table: Description of local operations.

Type	Includes	Example
Primary	Creation of a layer from nothing.	Rasters of constant value or containing randomly generated values.
Unary	Conversion of units of measurement and as intermediary steps of spatial analysis.	Rescaling, negation, comparing or applying mathematical functions reclassification.
Binary	Operations on ordered pairs of numbers in matching pixels between layers.	Arithmetic and logical combinations of rasters.
N-ary	Comparison of local statistics between several rasters (many to one or many to many).	Change or variety detection.

References

- What-is-cartography, resources: cca-acc.org, Retrieved 05, May 2020
- What-is-an-atlas-1435685: thoughtco.com, Retrieved 27, January 2020
- Globe-cartography, science: britannica.com, Retrieved 10, August 2020
- Concepts-of-Map-Algebra: dspmuranchi.ac.in, Retrieved 05, April 2020
- Map-algebra-global-zonal-focal-local: gisgeography.com, Retrieved 19, July 2020

GIS File Formats

A standard which is used to encode the geographical information into a computer file is known as a GIS file format. This chapter explains the two types of GIS file formats, GIS raster file formats and GIS vector file formats. All the diverse facts of GIS file formats have been carefully analyzed in this chapter.

File formats define in which way the geographical features are stored. Raster and vector data are stored in different formats.

There are numerous formats available for both raster and vector data. It is important to consider the file format of GIS data because software programs rarely support all file types. If you want to use GIS data that was saved in a particular format not supported by your GIS program, you must either find a way to transform the data or simply use another GIS program.

Nearly all GIS programs have their own file format. These file formats were created to optimise the efficiency of the program itself, and were not designed to be used in other external programs. Nevertheless, most GIS programs support other formats by having functions to import and export datasets. These functions are usually well documented and standardised.

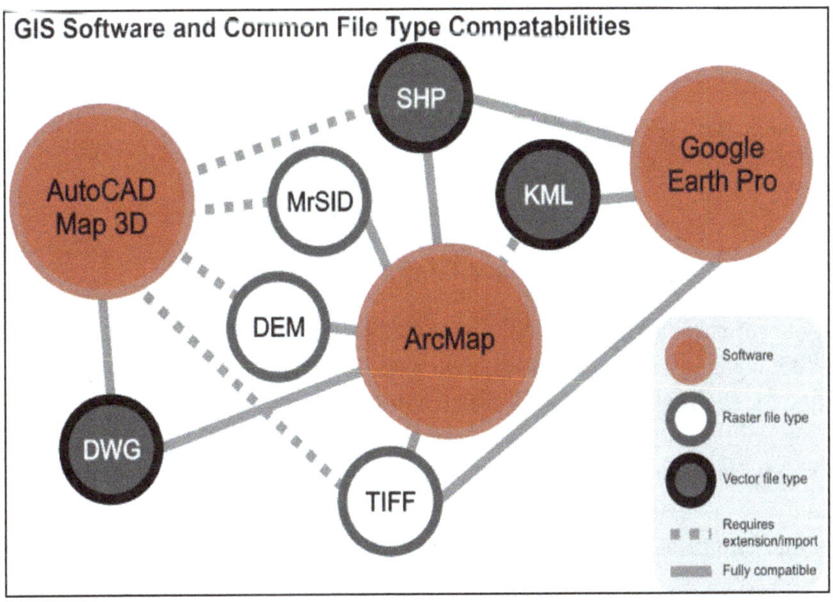

There are well over 100 GIS, CAD and Remote Sensing applications on the market. Listed here are some of the more common file formats along with associated file extension (the last 3 charcters after the period in the file name).

Raster

A raster data type is, in essence, any type of digital image represented by reducible and enlargeable grids. Anyone who is familiar with digital photography will recognize the Raster graphics pixel as the smallest individual grid unit building block of an image, usually not readily identified as an artifact shape until an image is produced on a very large scale. A combination of the pixels making up an image color formation scheme will compose details of an image, as is distinct from the commonly used points, lines, and polygon area location symbols of scalable vector graphics as the basis of the vector model of area attribute rendering. While a digital image is concerned with its output blending together its grid based details as an identifiable representation of reality, in a photograph or art image transferred into a computer, the raster data type will reflect a digitized abstraction of reality dealt with by grid populating tones or objects, quantities, cojoined or open boundaries, and map relief schemas. Aerial photos are one commonly used form of raster data, with one primary purpose in mind: to display a detailed image on a map area, or for the purposes of rendering its identifiable objects by digitization. Additional raster data sets used by a GIS will contain information regarding elevation, a digital elevation model, or reflectance of a particular wavelength of light, Landsat, or other electromagnetic spectrum indicators.

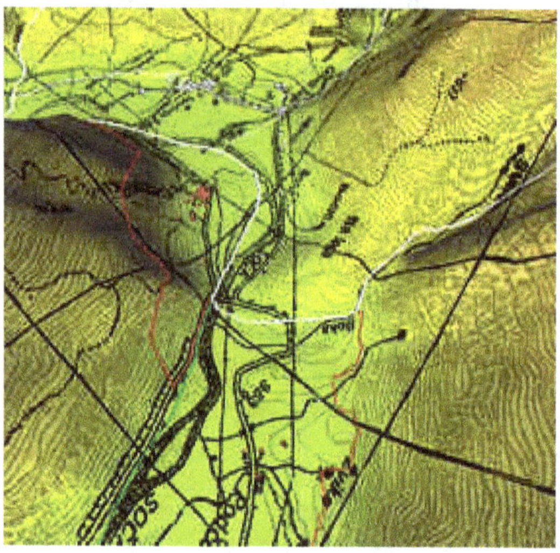

Digital elevation model, map (image), and vector data.

Raster data type consists of rows and columns of cells, with each cell storing a single value. Raster data can be images raster images with each pixel (or cell) containing a color value. Additional values recorded for each cell may be a discrete value, such as land use, a continuous value, such as temperature, or a null value if no data is available. While a raster cell stores a single value, it can be extended by using raster bands to represent RGB (red, green, blue) colors, colormaps (a mapping between a thematic code and RGB value), or an extended attribute table with one row for each unique cell value. The resolution of the raster data set is its cell width in ground units.

Raster data is stored in various formats; from a standard file-based structure of TIFF, JPEG, etc. to binary large object (BLOB) data stored directly in a relational database management system (RDBMS) similar to other vector-based feature classes. Database storage, when properly indexed, typically allows for quicker retrieval of the raster data but can require storage of millions of significantly sized records.

Vector

In a GIS, geographical features are often expressed as vectors, by considering those features as geometrical shapes. Different geographical features are expressed by different types of geometry:

- Points

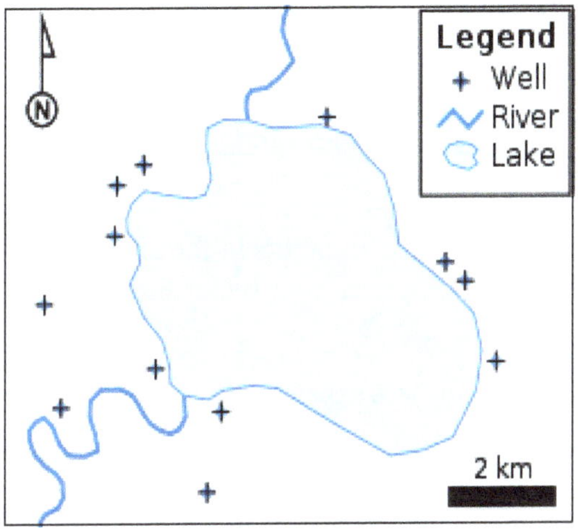

A simple vector map, using each of the vector elements: points for wells, lines for rivers, and a polygon for the lake.

  Zero-dimensional points are used for geographical features that can best be expressed by a single point reference—in other words, by simple location. Examples include wells, peaks, features of interest, and trailheads. Points convey the least amount of information of these file types. Points can also be used to represent areas when displayed at a small scale. For example, cities on a map of the world might be represented by points rather than polygons. No measurements are possible with point features.

- Lines or polylines

  One-dimensional lines or polylines are used for linear features such as rivers, roads, railroads, trails, and topographic lines. Again, as with point features, linear features displayed at a small scale will be represented as linear features rather than as a polygon. Line features can measure distance.

- Polygons

  Two-dimensional polygons are used for geographical features that cover a particular area of the earth's surface. Such features may include lakes, park boundaries, buildings, city boundaries, or land uses. Polygons convey the most amount of information of the file types. Polygon features can measure perimeter and area.

Each of these geometries are linked to a row in a database that describes their attributes. For example, a database that describes lakes may contain a lake's depth, water quality, pollution level. This information can be used to make a map to describe a particular attribute of the dataset. For example, lakes could be coloured depending on level of pollution. Different geometries can also be compared. For example, the GIS could be used to identify all wells (point geometry) that are within one kilometre of a lake (polygon geometry) that has a high level of pollution.

Vector features can be made to respect spatial integrity through the application of topology rules such as 'polygons must not overlap'. Vector data can also be used to represent continuously varying phenomena. Contour lines and triangulated irregular networks (TIN) are used to represent elevation or other continuously changing values. TINs record values at point locations, which are connected by lines to form an irregular mesh of triangles. The face of the triangles represent the terrain surface.

Advantages and Disadvantages

There are some important advantages and disadvantages to using a raster or vector data model to represent reality:

- Raster datasets record a value for all points in the area covered which may require more storage space than representing data in a vector format that can store data only where needed.
- Raster data is computationally less expensive to render than vector graphics.
- There are transparency and aliasing problems when overlaying multiple stacked pieces of raster images.
- Vector data allows for visually smooth and easy implementation of overlay operations, especially in terms of graphics and shape-driven information like maps, routes and custom fonts, which are more difficult with raster data.
- Vector data can be displayed as vector graphics used on traditional maps, whereas raster data will appear as an image that may have a blocky appearance for object boundaries. (depending on the resolution of the raster file).
- Vector data can be easier to register, scale, and re-project, which can simplify combining vector layers from different sources.
- Vector data is more compatible with relational database environments, where they can be part of a relational table as a normal column and processed using a multitude of operators.
- Vector file sizes are usually smaller than raster data, which can be tens, hundreds or more times larger than vector data (depending on resolution).
- Vector data is simpler to update and maintain, whereas a raster image will have to be completely reproduced. (Example: a new road is added).
- Vector data allows much more analysis capability, especially for "networks" such as roads, power, rail, telecommunications, etc. (Examples: Best route, largest port, airfields connected to two-lane highways). Raster data will not have all the characteristics of the features it displays.

Non-spatial Data

Additional non-spatial data can also be stored along with the spatial data represented by the coordinates of a vector geometry or the position of a raster cell. In vector data, the additional data contains attributes of the feature. For example, a forest inventory polygon may also have an identifier value and information about tree species. In raster data the cell value can store attribute information, but it can also be used as an identifier that can relate to records in another table.

Software is currently being developed to support spatial and non-spatial decision-making, with the solutions to spatial problems being integrated with solutions to non-spatial problems. The end result with these flexible spatial decision-making support systems (FSDSSs) is expected to be that non-experts will be able to use GIS, along with spatial criteria, and simply integrate their non-spatial criteria to view solutions to multi-criteria problems. This system is intended to assist decision-making.

Popular GIS File Formats

Raster Formats

- ADRG – National Geospatial-Intelligence Agency (NGA)'s ARC Digitized Raster Graphics.
- Binary file – An unformatted file consisting of raster data written in one of several data types, where multiple band are stored in BSQ (band sequential), BIP (band interleaved by pixel) or BIL (band interleaved by line). Georeferencing and other metadata are stored one or more sidecar files.
- Digital raster graphic (DRG) – digital scan of a paper USGS topographic map.
- ECRG – National Geospatial-Intelligence Agency (NGA)'s Enhanced Compressed ARC Raster Graphics (Better resolution than CADRG and no color loss).
- ECW – Enhanced Compressed Wavelet (from ERDAS). A compressed wavelet format, often lossy.
- Esri grid – proprietary binary and metadataless ASCII raster formats used by Esri.
- GeoTIFF – TIFF variant enriched with GIS relevant metadata.
- IMG – ERDAS IMAGINE image file format.
- JPEG2000 – Open-source raster format. A compressed format, allows both lossy and lossless compression.
- MrSID – Multi-Resolution Seamless Image Database (by Lizardtech). A compressed wavelet format, allows both lossy and lossless compression.
- netCDF-CF – netCDF file format with CF medata conventions for earth science data. Binary storage in open format with optional compression. Allows for direct web-access of subsets/ aggregations of maps through OPeNDAPprotocol.
- RPF – Raster Product Format, military file format specified in MIL-STD-2411.
    - ◊ CADRG – Compressed ADRG, developed by NGA, nominal compression of 55:1 over ADRG (type of Raster Product Format).
    - ◊ CIB – Controlled Image Base, developed by NGA (type of Raster Product Format).

Vector Formats

- AutoCAD DXF – contour elevation plots in AutoCAD DXF format (by Autodesk).
- Cartesian coordinate system (XYZ) – simple point cloud.
- Digital line graph (DLG) – a USGS format for vector data.
- Esri TIN - proprietary binary format for triangulated irregular network data used by Esri.
- Geography Markup Language (GML) – XML based open standard (by OpenGIS) for GIS data exchange.
- GeoJSON – a lightweight format based on JSON, used by many open source GIS packages.
- GeoMedia – Intergraph's Microsoft Access based format for spatial vector storage.
- ISFC – Intergraph's MicroStation based CAD solution attaching vector elements to a relational Microsoft Access database.
- Keyhole Markup Language (KML) – XML based open standard (by OpenGIS) for GIS data exchange.
- MapInfo TAB format – MapInfo's vector data format using TAB, DAT, ID and MAP files.
- National Transfer Format (NTF) – National Transfer Format (mostly used by the UK Ordnance Survey).
- Spatialite – is a spatial extension to SQLite, providing vector geodatabase functionality. It is similar to PostGIS, Oracle Spatial, and SQL Server with spatial extensions.
- Shapefile – a popular vector data GIS format, developed by Esri.
- Simple Features – Open Geospatial Consortium specification for vector data.
- SOSI – a spatial data format used for all public exchange of spatial data in Norway.
- Spatial Data File – Autodesk's high-performance geodatabase format, native to MapGuide.
- TIGER – Topologically Integrated Geographic Encoding and Referencing.
- Vector Product Format (VPF) – National Geospatial-Intelligence Agency(NGA)'s format of vectored data for large geographic databases.

Grid Formats (for Elevation)

- USGS DEM – The USGS' Digital Elevation Model.
- GTOPO30 – Large complete Earth elevation model at 30 arc seconds, delivered in the USGS DEM format.
- DTED – National Geospatial-Intelligence Agency (NGA)'s Digital Terrain Elevation Data, the military standard for elevation data.

- GeoTIFF – TIFF variant enriched with GIS relevant metadata.
- SDTS – The USGS' successor to DEM.

Other Formats

- Dual Independent Map Encoding (DIME) – A historic GIS file format, developed in the 1960s.
- Geographic Data Files (GDF) — An interchange file format for geographic data.
- GeoPackage (GPKG) – An standards-based open format based on the SQLite database format for both vector and raster data.
- Well-known text (WKT) – A text markup language for representing feature geometry, developed by Open Geospatial Consortium.
- Well-known binary (WKB) – Binary version of Well-known text.
- World file – Georeferencing a raster image file (e.g. JPEG, BMP).

GIS Raster File Formats

ECW (File Format)

ECW or the enhanced compressed wavelet file extension is a type of image file format compression wherein it basically stores information captured by the used of satellites and aerial cameras or gcoreaders placed in space. This file type is known to be used in capturing aerial pictures via satellites for the development of geolocation and mapping databases. ECW files are popular among developers and users of Geographic Information System and thus, there are a lot of applications that can be utilized to view files with.ecw format.

Earth Resource Mapping pioneered the use of enhanced compressed wavelet and later owned by ERDAS (Earth Resource Data Analysis System). ECW is an ideal image format for large image file because of its ability to achieve large compression rates that other formats have in difficulty.

ECW files can achieve compression rates between 1:10 and 1:100, making them ideal for compressing large images.

Applications that support the ECW file extension for images include Adobe Photoshop 4.0 and 7.0, AutoCAD, Bentley MicroStation, ERDAS IMAGINE and other software from the company, ArcGIS, and others.

The ECW extension is also used for waveset format files that store articulation and sample data used by the Ensoniq software MIDI synthesizer of sound cards equipped with specific chips. Those sound cards include the Ensoniq AudioPCI and various Sound Blaster cards from Creative Labs, which now owns E-MU Systems, the original developer of Ensoniq software.

Properties

Map projection information can be embedded into the ECW file format to support geospatial applications.

Image data of up to 65,535 bands (layers or colors) can be compressed into the ECW v2 or v3 file format at a rate of over 25 MB per second on an i7 740QM (4-cores) 1.731 GHz processor using v4.2 of the ECW/JP2 SDK. Data flow compression allows for compression of large images with small RAM requirements. The file format can achieve typical compression ratios from 1:2 to 1:100.

The ECW Protocol (ECWP) is an efficient streaming protocol used to transmit ECW and JPEG2000 images over networks, such as the Internet. ECWP supports ECWPS for private and secure encrypted streaming of image data over public networks such as the Internet.

There is a very fast read-only SDK supporting ECW and JPEG2000 which is available for no charge for desktop implementation for Windows, Linux and MacOSX. A read / write SDK can be purchased for desktop and server implementations for Windows, Linux and MacOSX). A full functioning server implementation (using ECW, JPEG2000, ECWP and JPIP) is available within the PROVIDER SUITE of the Power Portfolio (formerly IWS) license. A previous version of the SDK (3.3) is available in open source, and can be used for non-Microsoft operating systems, such as Linux, macOS or Android.

USGS DEM

The USGS DEM standard is a geospatial file format developed by the United States Geological Survey for storing a raster-based digital elevation model. It is an open standard, and is used throughout the world. It has been superseded by the USGS's own SDTS format but the format remains popular due to large numbers of legacy files, self-containment, relatively simple field structure and broad, mature software support.

DEM Level

A USGS DEM can be classified into one of four levels of quality. This is due to the multiple methods of data collection, and certainty in the data.

DEM level	DEM series	Production process	Comments
Level 1	7.5-minute 30-minute	Created by autocorrelation or manual profiling from aerial photographs	May be derived from a wide variety of sources using any of a number of USGS approved techniques and instrumentation
Level 2	All series	Created from digital line graph (DLG) contours or equivalent, from any USGS map series up to 1:100000 scale	The accuracy and data spacing are intended to support computer applications that analyze hypsographicfeatures to a level of detail similar to manual interpolations of information from printed source maps
Level 3	7.5-minute	Created from DLG that have vertically integrated with all categories of hypsography, hydrography, ridge line, break line, drain files, and all vertical and horizontal control network	Requires a system of logic to interpret the data

Level 4	All series	Created from electronic (non-photogrammetric) imaging sensor systems, either active (e.g., radar or lasar) or passive (received radiant energy)	The process includes all additional post-processing required to bring the DEM grid to its final form

Format Structure

The USGS DEM format is a self-contained (single file) set of ASCII-encoded (text) 1024-byte (1024 ASCII chars) blocks that fall into three record categories called A, B, and C. There is no cross-platform ambiguity since line ending control codes are not used, and all data including numbers is represented in readable text form. There is no known binary analogue of the format, although it is common practice to compress the files with gzip.

Floating-point numbers are encoded using Fortran scientific notation, so C/C++programs need to swap the "D" exponent-indicating character with "E" when parsing (and vice versa when writing).

A record fields hold the origin, type, summary statistics and the measurement systems used by the profiles. The A record appears once as the file's header, the C record also appears once as the trailer, and multiple B records (called profiles) comprise the elevation data. A and C records each fit within one block, but a single B record typically requires multiple blocks. When such block-spanning occurs, data are shifted to start cleanly on each block boundary. A records also come in "old" and "new" flavors, because the USGS added several fields to the A record. One of the key items is the quadrangle, which is a set of four terrestrial coordinates describing the four-sided polygon enclosing the area of interest.

Starting Col	Ending Col	Description
0	134	Descriptive Name of the represented area
150	155	?
156	161	?
162	167	UTM Zone number
529	534	Unit of resolution of ground grid (0=radian;1=feet;2=metre;3=arc-second)
535	540	Unit of resolution Elevation (1=feet;2=metre)
546	569	Easting of the South West corner
570	593	Northing of the South West corner
594	617	Easting of the North West corner
618	641	Northing of the North West corner
642	665	Easting of the North East corner
666	689	Northing of the North East corner
690	713	Easting of the South East corner
714	737	Northing of the South East corner
738	761	Minimum elevation found in this file
762	786	Maximum elevation found in this file
816	827	Resolution per grid cell East – West
828	839	Resolution per grid cell North – South
858	863	Number of columns

B records (profiles) are a variable-length longitudinal column of raster elevations that start at a specified location. They are some multiple of 1024 bytes long and contain a small header summarizing the profile. The elevations are contiguous; breaks or other discontinuities are expressed using "void" elevations of value -32767. Each elevation is described as a six-character readable integer occupying a fixed location in a block. The profile header only appears in the first block, so subsequent blocks hold more elevation values. When reading the DEM file from first byte to last, one reads the profiles as columns from west to east. The elevations within a profile run from south to north. The variable-location and variable-length nature of profiles stems mainly from the use of the UTM (Universal Transverse Mercator) ground reference system. Since measurements within UTM employ fixed distances (e.g., 30 meters between elevation samples), the quadrangle must slightly distort to map such locations onto the spherical Earth. This distortion usually manifests as a rotated square, hence the elevation columns near the east and west edges start more northward and contain fewer samples.

C records contain root-mean squared error (RMSE) quality control data, using ten six-character integer fields.

MrSID

MrSID (Multi-resolution Seamless Image Database) is an image compressor, viewer, and file format for extremely large raster graphics images. MrSID, created by LizardTech, works by putting together hundreds of small image tiles into one large seamless image that can be compressed and decompressed with little or no degradation.

One problem previously associated with raster images was the long amount of time it took to open very large files. The creators of MrSID overcame that problem by giving the user the ability to decompress just the portion of the image they wanted to view. The user can move from one part of the image to another quickly without having to wait for the entire image to decompress.

Another problem associated with massive raster images was the large amount of storage space they required. MrSID uses a patented "wavelet" technology first developed at the National Research Laboratories at Los Alamos to achieve a compression ratio of 20:1 for greyscale images and 50:1 for full color images. Now, a satellite photo from space that once required the storage space of forty CD-ROMs, could be compressed with MrSID and stored on one CD-ROM.

The wavelet technology used by MrSID relies on advanced mathematical algorithms both to compress an image and build the viewer. Since the image remains geometrically accurate after compression, it can be geo-referenced before compression or overlaid with other referencing data. Because MrSID (.sid) files are a binary MIME type, they can be served over the Internet.

Common Uses

Fingerprints

According to the Open Source Geospatial Foundation (which releases GDAL), MrSID was developed "under the aegis of the U.S. government for storing fingerprints for the FBI."

Other Uses

In a 1996 entry for the R&D 100 Awards, LANL identified other uses for the format: "it can be used as an efficient method for storing and retrieving photographic archives; it can store and retrieve satellite data for consumer games and educational CD-ROMs; and it is well suited for use in vehicle navigation systems. Moreover, MrSID holds promise for being used in image compression and editing for desktop publishing and nonlinear digital video software."

For certain downloadable images (such as maps), American Memory at the Library of Congress began using MrSID in 1996; in January 2005 it also began using JPEG 2000. Depending on image content and color depth, compression of American Memory maps is typically better with MrSID, which on average achieves a compression ratio of approximately 22:1 versus the 20:1 achieved with JPEG 2000.

Software

LizardTech offers a software package called GeoExpress to read and write MrSID files. They also provide a free web browser plug-in for the Microsoft Windows operating system. (A Macintosh OS version of this viewer, introduced in 2005, was discontinued.) Most commercial GIS software packages can read some versions of MrSID files including those from GE Smallworld, ESRI, Intergraph, Bentley Systems, MapInfo, Safe Software, Autodesk, with ERDAS IMAGINE being able to both read and write MrSID files. GeoExpress can also generate JPEG 2000 (.jp2) data. When combined with LizardTech's Express Server,.sid and.jp2 data can be served quickly to a variety of GIS applications and other client applications either through direct integrations or via WMS.

There is no open source implementation of the MrSID format. Some open source GIS systems can read MrSID files, including MapWindow GIS and those based on GDAL. The Decode Software Development Kit (SDK) is made available as a free download from developer.lizardtech.com. This enables the capability to implement MrSID reading capability in any application.

Some image editing and management software systems can also read MrSID files, including XnView and IrfanView.

Technology

Encoding

MrSID technology uses lossless wavelet compression to create an initial image. Then the encoder divides the image into zoom levels, subbands, subblocks and bitplanes. After the initial encoding, the image creator can apply zero or more optimizations. While 2:1 compression ratios may be achieved losslessly, higher compression rates are lossy much like JPEG-compressed data.

Decoding

MrSID uses selective decoding meaning that the decoder does not have to decode the entire file to view a specific zoom level, image quality or scene for example.

Esri Grid

A grid is a raster data storage format native to Esri. There are two types of grids: integer and floating point. Use integer grids to represent discrete data and floating-point grids to represent continuous data.

Attributes for an integer grid are stored in a value attribute table (VAT). A VAT has one record for each unique value in the grid. The record stores the unique value (VALUE is an integer that represents a particular class or grouping of cells) and the number of cells (COUNT) in the grid represented by that value. For example, if 50 cells have a value of 1 representing a forest, then the VAT would show a VALUE = 1 and COUNT = 50 for each of the 50 cells.

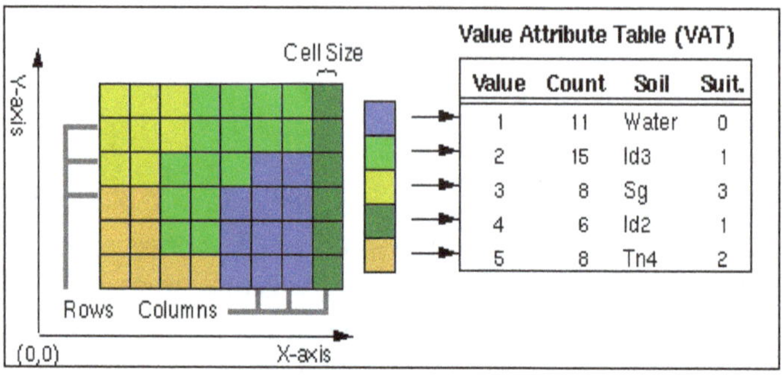

Floating-point grids do not have a VAT because the cells in the grid can assume any value within a given range of values. The cells in this type of grid do not fall neatly into discrete categories. The cell value itself is the attribute that describes the location. For example, in a grid that represents elevation data in meters above sea level, a cell with a value of 10.1662 indicates that the location is about 10 meters above sea level.

The range of data values that can be stored as grid values are as follows:

- Floating-point grids can store values from -3.4 x 1038 to 3.4 x 1038.
- Integer grids can store values from -2147483648 to 2147483647 (-231 to 231-1).

For integer grids, this information applies only to the VALUE item. An integer grid may have other INFO items added to its VAT whose range of values depends on the item definition.

The coordinate system of a grid is the same as that of other geographic data. The rows and columns are parallel to the x- and y-axes of the coordinate system. Since each cell within a grid has the same dimension as other cells, the location and area covered by any cell is easily determined by its row and column. The coordinate system of a grid is thus defined by the cell size, the number of rows and columns, and the x,y coordinate of the upper left corner. Grids also carry additional information, such as the coordinate system associated with the grid.

File Formats

ESRI ArcInfo ASCII Grid

Output format created by the GRIDASCII command (from ArcInfo Workstation) and by the Raster to ASCII tool in ArcGIS for Desktop. ESRI_ASCII_grid is an interchange format, used primarily for exchange with other programs. The file contents represents a single measure for each cell in a rectangular grid. Any coordinate system can be used to define the grid, including the widely used angular measurements with arc-seconds as units. Cells must be "square" in the chosen unit since cell size is defined by a single parameter.

The file begins with header information that defines the properties of the raster, such as the cell size, the number of rows and columns, and the coordinates of the origin of the raster. The header keywords can be in upper or lower case. The header information is followed by cell data; numeric values are ordered by column within row. Within each row, values are delimited by spaces and rows are separated by a carriage return. Either integer or floating point numbers can be used for the individual data values. The spatial location of the raster is specified by the location of the lower left cell, and by either the center of the lower left cell or the lower left corner of the lower left cell. Row 1 of the data is at the top of the raster, row 2 is below row 1, and so on.

- NCOLS *xxx*
- NROWS *xxx*
- XLLCENTER *xxx* or XLLCORNER *xxx*
- YLLCENTER *xxx* or YLLCORNER *xxx*
- CELLSIZE *xxx*

  [the length of one side of a square grid cell in same units as coordinate system used to locate the lower left center]
- NODATA_VALUE *xxx*
- Row 1 data
- Row 2 data
- etc.

Cells without actual values can be assigned a *NoData* value. The default value for *NoData* is -9999 but a different value can be specified in the header as shown in example below.

A short example:

- NCOLS 4
- NROWS 6
- XLLCORNER 378923
- YLLCORNER 4072345
- CELLSIZE 30
- NODATA_VALUE -99
- -99 -99 5 2
- -99 20 100 36

- 3 8 35 10
- 32 42 50 6
- 88 75 27 9
- 13 5 1 -99

Various discussions suggest that scientific notation may be used for the numeric values.

ESRI ArcInfo Grid

ESRI ArcInfo Grid (ESRI_grid), also known as ArcGrid, is a raster file format developed by ESRI to contain information about geographic space in a grid. A grid defines geographic space as an array of equally sized square grid points (also known as cells) arranged in rows and columns. Grids are useful for representing geographic phenomena that vary continuously over space, and for performing spatial modeling and analysis of flows, trends, and surfaces such as hydrology. Each grid cell, referenced by its x,y coordinate location, stores a numeric value that represents a geographic attribute (such as elevation or surface slope) for that unit of space. In the ESRI_grid format this numeric attribute may be a 32-bit signed integer or a 32-bit (single precision) floating point number.

The native binary ESRI Grid format comprises a number of component files. These files are stored within an ArcInfo workspace, which may incorporate several grids and other geospatial datasets, such as coverages. An ArcInfo workspace contains one INFO subdirectory and a separate subdirectory for each ESRI_grid. The INFO directory holds several files that maintain information about each grid and other datasets in the workspace. The name of an ESRI_grid cannot start with a number or use spaces, and cannot be longer than 13 characters (a multiband grid is allowed up to 9 characters). Each GRID subdirectory contains several component files that store geographic location and attribute data for the corresponding grid. Not all of them are necessary for the ESRI_grid to display. Each component file is stored in one of three storage types: INFO format ASCII format, or binary format.

In a GRID directory, the following tables and files are found:

- dblbnd.adf (aka bnd.adf) -- The boundary file contains the minimum and maximum x,y coordinates for an ESRI_grid. The boundary is a rectangle that encompasses the cells of a grid; it is stored in map coordinates. All grid BNDs are stored in double precision. This file will have a corresponding.nit and.data files in the INFO directory of an ESRI_grid. This information is stored in INFO format.
- hdr.adf -- The header file contains information about the ESRI-grid's cell resolution, type (integer or floating point), compression, and tile information. This information is in binary format.
- log -- The log file contains a history about the creation of and alterations to an ESRI_grid. This information is in ASCII format.
- prj.adf -- The projection file stores the known map projection information of the ESRI_grid. It is optional. This information is in ASCII format.

- sta.adf -- The statistics (minimum, maximum, mean, and standard deviation) of the cell values are stored in the sta.adf file. Values in the statistics are floating point values. They should be not changed using INFO commands. This file will have a corresponding.nit and. data files in the INFO directory of an ESRI_grid. More information about sta.adf files can be found in What is the file structure of an Arc/INFO Grid? This information is in INFO format.
- vat.adf -- The value attribute table (VAT) stores the the attribute data associated with the zones of the grid. The vat.adf can only exist for integer grids. This file will have a corresponding.nit and.data files in the INFO directory of an ESRI_grid. More information about vat.adf files can be found in What is the file structure of an Arc/INFO Grid? This information is in INFO format.
- w001001.adf -- Stores the value of each cell in the first (base) tile of the grid. Subsequent tiles will be numbered sequentially. This information is in binary format.
- w001001x.adf -- Stores the index for the blocks in the tile. If the grid has more than one tile, the value and index.adf file names will increment correspondingly. This information is in binary format.
- Also included in an ArcInfo Workspace is an INFO subdirectory containing files that maintain information about the ESRI_grid and other associated datasets in the workspace. In an INFO directory, the following tables and files are found:
- arcNNNN.dat (where N is any integer) -- Stores the relative path to the grid data file (e.g., "../soils/vat.adf"). This information is stored in INFO format.
- arcNNNN.nit -- Stores file structure and field definition information of the grid. For any particular grid, the NNNN values for.dat and.nit should match. This information is stored in INFO format.
- arc.dir -- Contains the data dictionary for the grid.

Two types of grids can be found in the ESRI_grid, integer and floating point. Integer grids are used to represent discrete data, i.e., that which has known and definable boundaries. An integer grid is used to determine where a data object begins and where it ends, e.g., a lake in a landscape. Floating point grids represent continuous data, i.e., data whose boundaries cannot be precisely defined; instead, the boundaries are measured in terms of a relationship to a fixed point such as elevation measured from sea level.

Grids are implemented using a tiled raster data structure in which the basic unit of data storage is a rectangular block of cells. Blocks are stored on disk in compressed form in a variable-length file structure referred to as a tile. Each block is stored as one variable-length record.

The size of the tile for a grid is based on the number of rows and columns in the grid at the time of creation. The upper limit on the size of a tile is set by the application and is very large (currently set at 4,000,000 x 4,000,000 cells). As a result, most grids used for GIS applications are automatically stored in a single tile. The spatial data for a grid is automatically split across multiple tiles if the size of the grid at the time of creation is larger than the upper limit for the size of a tile.

Grids can be combined into an ordered set of spatially overlapping grids treated as a single entity, called a grid *stack*. Grids (also known as "layers") are combined into a stack to facilitate multivariate analysis such as cluster analysis, classification, and principal component analysis. A stack has the following characteristics: a set of layers with each layer corresponding to a grid, a map extent, a cell size, a data type, and a projection stored in an ASCII file. Each layer in the grid has an index number indicating its order in the stack, all of which must be in the same workspace. The boundaries of the input layers for a grid stack can overlap exactly, partially, or not all, but the stack is considered to be only the area where the layers overlap. Computations against the stack will only be done against the overlapping area as declared by the intersection of the boundaries in the bnd.adr file. If no common area exists between the input layers, the stack will be empty and no computations will occur. As with a simple grid, the name of a grid stack cannot start with numbers, contain spaces or be longer than nine characters.

Any number of data types (real or integer) within a layer can be combined into a stack. The cell size of stack defaults to the coarsest layer in the stack. The stack is stored in the INFO directory as is information about the grids of which it is comprised. The grids of which a stack is comprised are not duplicated in a grid, but instead are referenced in the stack-related files. The directory in which the stack is located will have the following files within it:

- *.stk -- (where * indicates the user generated name of the stack) This file contains a table that stores the names of the grids that comprise the stack, their corresponding index values, and the record number of the grid. The INDEX column gives the position of a grid in the stack while the GRID column lists the grid names that comprise the stack. Other columns can be added to the stack table, but the INDEX and GRID values should not be altered unless using appropriate stack management commands found in the GRID functionality within the ArcGIS tools.

- prj.adr -- As with the description for prj.adr above, this file, when present, stores the projection information of the stack. Since a stack is treated as a single entity, all grids in a stack must be in the same projection. If the projection is unknown for all input grids in the stack, no prj.adr file is created. The projection information is used to ensure that the same geographic area is being occupied by each grid in the stack.

Grid Data Storage

A grid is stored in an ArcInfo workspace. The grid, like a coverage, is stored as a separate directory with associated tables and files that contain specific information about the grid. In an integer grid directory (originally created by ArcInfo Workstation), the following tables and files are found: the BND table, which stores the boundary of the grid; the HDR file, which stores specific information describing the grid, for example, cell resolution and blocking factor; the STA table, which contains statistics for the grid; the VAT table, which stores the attribute data associated with the zones of the grid; the log file (LOG), which monitors the activity that has occurred on the grid; and the tile file w001001.adf (q0x1y1), which stores the cell data and the accompanying index file w001001x.adf (q0x1y1x) that indexes the blocks in the tile and the LOG. (Some of these, such as the log file, may not exist if created using ArcGIS operators).

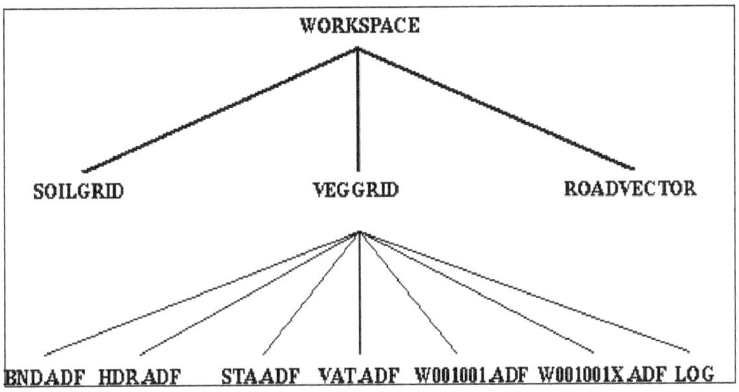

If a grid is altered, the values and information contained in the files and tables are updated immediately. The information contained in an INFO table is accessible to the user and provides information about the grid.

BND Table

A grid BND contains the boundary of the grid. The boundary is a rectangle that encompasses the cells of a grid; it is stored in map coordinates. All grid BNDs are stored in double precision.

The minimum coordinates in the BND table are for the lower left corner of the lower left cell in the grid. The maximum coordinates are for the upper right corner of the upper right cell in the grid.

HDR File

The HDR is a binary file. Information stored in the file includes the cell size, type of grid (integer or floating point), compression technique, blocking factor, and tile information.

STA Table

The STA table is an INFO table that contains statistical data about a grid. The minimum, maximum, mean, and standard deviation for the grid are stored as floating-point values in the STA table. You should not attempt to alter these values directly.

Because NoData represents an unknown value, NoData is not used in calculating the statistics in the STA table.

When a bilevel grid (containing only 0 and 1 values) is created, the STA table contains the value 0 for the mean and -1 for standard deviation. The standard deviation value -1 indicates that statistics have not been calculated for a grid.

A standard deviation value of -2 indicates that the grid contains only NoData cells.

VAT Table

The VAT is an INFO table that stores attributes associated with the zones of a grid. Only integer grids have a VAT associated with them. Every VAT has at least two items, VALUE and COUNT. The VALUE item contains integer values that are used to distinguish the characteristics of one location from the other locations in a grid. All cells that are assigned the same value contain the same characteristics and, therefore, belong to the same zone. COUNT is the number of cells in a zone.

New items can be added to the VAT. The VALUE and COUNT items should not be changed, and the VAT must be kept sorted on the VALUE item. Never add new items before VALUE or COUNT.

Cells containing NoData are not represented in the VAT.

Below is an example of a VAT:

Record	VALUE	COUNT
1	0	628872
2	1	265043
3	2	151150
4	3	3185652
5	4	79983
6	5	4782
7	6	74334
8	7	8877
9	8	1817
10	9	491
11	10	858
12	11	8770
13	12	28789
14	13	72539
15	14	3686
16	15	3932
17	16	13227
18	17	1890
19	18	1305
20	19	427286
21	20	6695

Tile Files

The w001001.adf (q0x1y1) and w001001x.adf (q0x1y1x) files store the data and the index for the first, or base tile, in a grid. The upper limit on the size of a tile is very large, and most grids are stored using a single tile. If additional tiles are used, they are automatically numbered based on their spatial relationship to the first tile. Tiles are implemented as variable-length binary files. With versions prior to ARC/INFO 7.x, these files were named q0x1y1 and q0x1y1x and still work with the current software.

LOG File

The LOG file is an ASCII file that contains information about the creation of and alterations to a grid. The LOG monitors the actions performed on the grid, but it does not contain every action performed with the grid. Since all Grid functions result in a new grid, only Grid commands, such as RENAME and COPY, can alter an existing grid and be entered into the LOG file. The LOG file can be accessed, like all ASCII files, through system commands or any text editor.

Storage Limitations

- The name of a grid is limited as follows:
- It cannot be stored using spaces.
- It cannot start with a number.
- It cannot be longer than 13 characters (a multiband grid is allowed up to 9 characters).

There is a limit to the number of files that can be stored in an INFO directory for both coverages and grids. This total is approximately 10,000. Therefore, this limits the number of grids you can store in a workspace. For example, the following lists the theoretical maximum number of grid datasets that can be stored in a single workspace directory:

- Fewer than 5,000 floating point grids, or
- Fewer than 3,333 integer grids, with VATs (fewer than 5,000 if no VATs), or
- Fewer than 10,000 grid stacks

The preceding numbers are the theoretical maximums. If you have a process that will create interim grids (and therefore files in the INFO directory) these numbers will be less. Additionally, if you are storing a mix of files, such as grids and coverages, you will store fewer.

These numbers relate to the number of files in the grid folder that store information in the INFO directory. The limit is 10,000 (9,999), but it's not the total number of files in an INFO directory, it's the number of files pointing to the files in the INFO directory. For each grid, there are two files in the grid's folder pointing to files in the INFO folder: the BND (boundary) files, and STA table (statistics) files (9999/2≈5000). When a grid has a VAT, this also points to files in the INFO directory, so the number that can be stored is reduced again (9999/3≈3333). A grid stack only has a single file which points to the INFO directory (9999/1≈9999).

Stack

A stack consists of an ordered set of spatially overlapping grids (layers) treated as a single entity for multivariate analysis. Cluster analysis, classification, and principal component analysis all work on the layers in a stack.

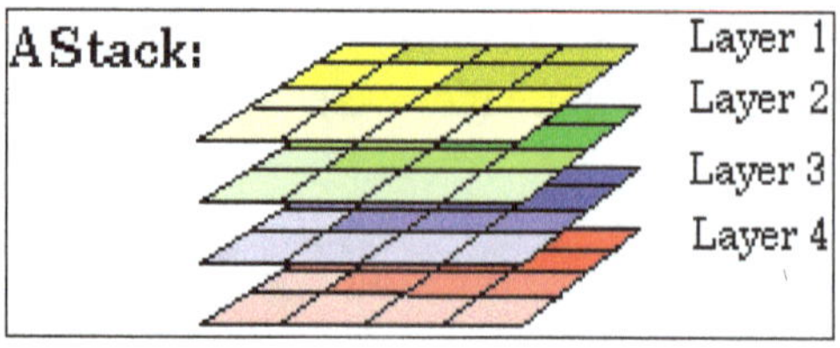

A stack has the following characteristics:

- A set of layers with each layer corresponding to a grid
- A map extent, or BND
- A cell size
- A data type
- A projection

Each layer specified in a stack has an index number indicating its order in the stack. The grids that make up a stack must be in the same workspace.

The boundaries of the input layers can overlap exactly, partially, or not at all, but only the area where layers overlap comprises the stack. The stack's BND is where the boundaries of its layers intersect. The computations of a multivariate analysis function occur on the overlapping area. If there is no common area between the input layers, the stack is empty and no computations occur.

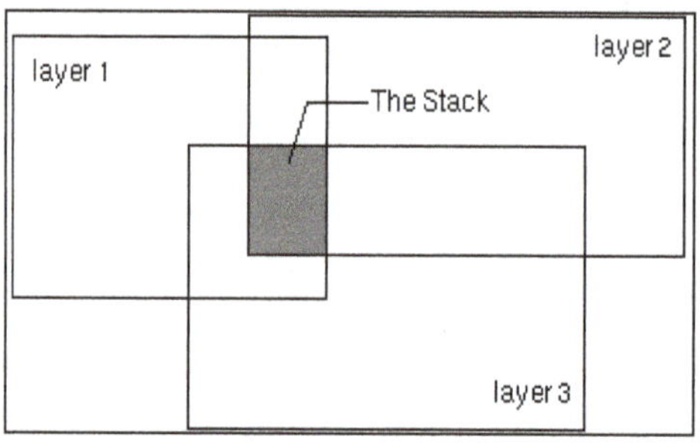

The cell size of a stack defaults to the coarsest layer in the stack.

You can combine any number of data types (real or integer) of the input grids in a stack. However, before applying a multivariate technique, you should be aware of what the values represent, whether categorical or continuous data, and the range or relative range of the values. In certain analysis, the input data type of the stack determines the data type of the output.

Projection information associated with the input grids is stored with the stack. Since a stack is treated as a single entity, all grids in a stack must be in the same projection. The projection information is used to ensure that each grid of the stack occupies the same geographic area.

Storing a Grid Stack

A stack is stored in a directory structure similar to a grid or coverage. There are two files in the stack directory: an external INFO STK table and an ASCII PRJ file. The actual grids that comprise the stack are not stored in the stack. They are ordinary grids in your workspace. That means any grid can be used in more than one stack. The STK table stores the names of the grids that comprise the stack and their corresponding index values:

GRID: LIST JER135.STK

Record	INDEX	GRID
1	1	jer1
2	2	jer3
3	3	jer5

The INDEX item gives the position of a grid in the stack, while the GRID item lists the grid names that comprise the stack. The spatial data of the input grids is not duplicated in the stack. As a result, the stack always reflects the latest version of the input grids. The STK file is as accessible as any other INFO file. You can add items for descriptive purposes, such as an item for storing the date that the data was collected, but don't use INFO to alter the values in the INDEX item or names in the GRID item. All manipulations to these items should only be performed using a variety of the stack management commands available in Grid.

The PRJ file, when present, stores the projection information of the stack:

Projection STATEPLANE

Zone	4701
Datum	NAD27
Zunits	NO
Units	FEET
Spheroid	CLARKE1866
Xshift	0.0000000000
Yshift	0.0000000000
Parameters	

If the projection is unknown for all input grids in the stack, no PRJ file is created.

The name of a grid stack cannot be stored using spaces, cannot start with a number, and cannot be longer than 9 characters.

NoData in a Grid

Every cell in a grid has a value assigned to it; however, cells without actual values can be assigned NoData on the grid representing that theme. NoData and 0 (zero) are not the same; 0 is a valid value. For this reason, NoData cells cannot be used in calculating the statistics in a grid's STA table.

GIS Vector File Formats

CityGML

CityGML is a common information model for representing 3D urban objects. It defines classes and relations for the most relevant topographic objects in cities and regional models with respect to their geometric, topological, semantic and appearance properties. "City" is broadly defined to include not just built structures, but also elevation, vegetation, water bodies, "sidewalk furniture" and more. Included are generalization hierarchies between thematic classes, aggregations, relationships between objects and spatial properties.

These thematic information types go beyond graphic exchange formats and allow users to employ virtual 3D city models for sophisticated analysis tasks in different application domains such as simulation, urban data mining, facilities management, decision support and thematic inquiries.

CityGML is an open data model and XML-based format for storing and exchanging virtual 3D city models. It is implemented as an application schema of GML3, the extensible international standard for spatial data exchange developed within the Open Geospatial Consortium (OGC) and ISO TC211.

GML3, used with other OGC standards - mainly the OpenGIS Web Feature Service (WFS) Specification - provides a framework for exchange of simple and complex 3D models. However, WFS and GML3 only establish syntactic interoperability. A GML3 document needs to be structured by the definition of an application schema that is tailored to a specific application domain. In this case, the application domain is 3D city modeling and the GML3 application schema (or profile) is CityGML.

CityGML takes advantage of other open standards and its development has proceeded in careful cooperation with other groups. For example, graphic rendering of data encoded in CityGML can be accomplished using standardized computer graphics data formats like VRML, GeoVRML, X3D or Universal 3D (U3D). To ensure European acceptance, the developers of CityGML have coordinated with EuroSDR, a spatial data research organization consisting of delegates from geographic information production organizations and research centers from 18 member states in Europe, together with participants from industry and the commercial sector.

Within the OGC, discussion of CityGML takes place within the OGC CAD/GIS/3D Interoperability Working Group. CityGML will be submitted to the OGC Technical Committee, meeting in Edinburgh, Scotland during the week of June 26-29, as a candidate OpenGIS specification for review and possible adoption by the membership. The intention is for it to become an open standard and therefore to be used free of charge.

Development of CityGML began in 2002 by the members of the Special Interest Group 3D (SIG 3D) of the Geodata Infrastructure North-Rhine Westphalia (GDI NRW) initiative in Germany. The GDI NRW has been active in the OGC since 1999. The SIG 3D is an open group consisting of more than 70 companies, municipalities, and research institutions working on developing and commercially exploiting interoperable 3D models and visualization. Another product of SIG 3D's activities is a proposed Web 3D Service (W3DS) standard, a 3D portrayal service that currently has the status of a discussion paper in the OGC (OGC Doc. No. 05-002).

Some of the entities working on and supporting CityGML are the municipalities of Berlin, Hamburg, Cologne, Düsseldorf, Recklinghausen and Leverkusen; the Ordnance Survey Great Britain; the State Mapping Agency of North-Rhine Westphalia; companies including T-Mobile, Rheinmetall Defence Electronics, Snowflake, CPA Geo-Information, GIStec and 3D Geo; and research institutions including the universities of Bonn, Potsdam, Dortmund, Applied Sciences Stuttgart, the Helmholtz Research Centre Karlsruhe, and the Fraunhofer Institute for Graphics Research in Darmstadt.

CityGML Features

CityGML represents the graphic appearance of city models and also the semantic or thematic properties, taxonomies and aggregations of digital terrain models, sites (including buildings, bridges and tunnels), vegetation, water bodies, transportation facilities, etc. The underlying model differentiates five consecutive levels of detail (LoD), where objects become more detailed with increasing LoD, both in geometry and thematic differentiation. CityGML files can - but don't have to - contain multiple representations for each object in different LoD simultaneously.

At this point, the UML diagrams, XML schema definitions, example datasets and free viewer applications are available from the CityGML homepage. Basic concepts and most of the modeling decisions are explained in a number of conference papers and presentation slides which can also be found on the website. The OGC will publish the CityGML specification document when it becomes an approved Discussion Paper.

An open source GML3 3D viewer application called Aristoteles has been developed by the Institute for Cartography and Geoinformation at the University of Bonn. It is based on Java 1.5 and Java3D. Aristoteles is not restricted to CityGML files; it can visualize arbitrary GML3 files (like the German cadastre standard ALKIS/NAS or the OS Mastermap from UK).

A free CityGML viewer application called LandXPlorer was developed by 3DGeo and the Hasso-Plattner-Institute (HPI) of the University of Potsdam. The LandXplorer CityGML Tool allows users to interactively view, import and export CityGML-based model data. For non-profit use and with limitations regarding the model size and the available functions, the LandXplorer CityGML Tool can be used for free.

CityGML in the OGC OWS-4 Testbed

The OGC's main interoperability initiative this year is the fourth OGC Web Services testbed (OWS-4). OWS-4 kicked off in June and will conclude with an interoperability demonstration in the fall of 2006. OWS-4 sponsors agreed that this initiative would advance geospatial standards in the area of CAD/GIS/BIM (building information model) integration, as well as in other areas such as sensor webs, geoprocessing workflow and geospatial digital rights management.

OWS-4 is based on a fictitious scenario that will be played out in the final demonstration. The CAD/GIS/BIM use case in the scenario involves siting and planning a temporary field hospital in an emergency situation. The commander determines that a desirable potential site is an abandoned building at the local airport. Facility planners analyze the location, including consideration of access by road and helicopter. Security is a high concern.

CityGML will be rigorously exercised and perhaps extended by the OWS-4 participants. A CAD Model Server (CMS) will be created to provide access to various kinds of CAD data, such as as-built data for buildings originally in CAD files without coordinate system information, road data in civil design applications, underground utilities CAD drawings and manholes in photogrammetric, planimetric data in a CAD file.

A Web Feature Service (implementing the OGC WFS standard) will be developed to provide client programs with access to GIS site data (terrain, orthophotos, land use), accessible as GML, and to Aeronautical Information Exchange Model (AIXM) data available for a helicopter pad.

The scenario outlines a basic course of action that will put the various servers and clients through a series of operations designed to test and demonstrate the utility of CityGML.

A virtual reality constructed from diverse kinds of data representing the built environment can benefit millions of people working in a long list of industries. As bandwidth, processing and storage keep getting faster and cheaper, lowering technical and financial hurdles, it makes more and more sense for stakeholder groups to imagine how such a shared virtual reality might benefit them.

CityGML is the result of a well-organized effort to bring together related standards through careful coordination with various standards groups and user groups. After years of development, it is now in "beta" and it will be tested in OWS-4. CityGML deserves careful review by a wide range of industry groups studying geospatial data sharing, CAD/GIS interoperability, industry foundation classes (IFCs), building information models (BIMs), and workflow and business models in building-related professions.

GeoJSON

GeoJSON is an open standard format designed for representing simple geographical features, along with their non-spatial attributes. It is based on JSON, the JavaScript Object Notation.

The features include points (therefore addresses and locations), line strings (therefore streets, highways and boundaries), polygons (countries, provinces, tracts of land), and multi-part collections of these types. GeoJSON features need not represent entities of the physical world only; mobile routing and navigation apps, for example, might describe their service coverage using GeoJSON.

The GeoJSON format differs from other GIS standards in that it was written and is maintained not by a formal standards organization, but by an Internet working group of developers.

A notable offspring of GeoJSON is TopoJSON, an extension of GeoJSON that encodes geospatial topology and that typically provides smaller file sizes.

Example

```
{
  "type": "FeatureCollection",
  "features": [
   {
     "type": "Feature",
     "geometry": {
        "type": "Point",
        "coordinates": [102.0, 0.5]
   },
   "properties": {
     "prop0": "value0"
   }
  },
  {
   "type": "Feature",
   "geometry": {
     "type": "LineString",
     "coordinates": [
```

```
        [102.0, 0.0], [103.0, 1.0], [104.0, 0.0], [105.0, 1.0]
    ]
   },
   "properties": {
    "prop0": "value0",
    "prop1": 0.0
    }
   },
   {
    "type": "Feature",
    "geometry": {
      "type": "Polygon",
      "coordinates": [
      [
         [100.0, 0.0], [101.0, 0.0], [101.0, 1.0],
         [100.0, 1.0], [100.0, 0.0]
      ]
    ]
   },
   "properties": {
    "prop0": "value0",
    "prop1": { "this": "that" }
   }
  }
 ]
}
```

Supported Software

GeoJSON is supported by numerous mapping and GIS software packages, including OpenLayers, Leaflet, MapServer, Geoforge software, GeoServer, GeoDjango, GDAL, Safe Software FME, and

CartoDB. It is also possible to use GeoJSON with PostGIS and Mapnik, both of which handle the format via the GDAL OGR conversion library. Bing Maps, Yahoo!, HERE, and Google also support GeoJSON in their API services.

The Google Maps JavaScript API v3 directly supports the integration of GeoJSON data layers as of March 19, 2014. For the Julia language a GeoJSON.jl package is available.

GitHub also supports GeoJSON rendering and Potrace GeoJSON export.

Geojson.io supports GeoJSON rendering and editing in the web browser.

TopoJSON

TopoJSON is an extension of GeoJSON that encodes topology. Rather than representing geometries discretely, geometries in TopoJSON files are stitched together from shared line segments called *arcs*. *Arcs* are sequences of points, while line strings and polygons are defined as sequences of arcs. Each arc is defined only once, but can be referenced several times by different shapes, thus reducing redundancy and decreasing the file size. In addition, TopoJSON facilitates applications that use topology, such as topology-preserving shape simplification, automatic map coloring, and cartograms.

A reference implementation of the TopoJSON specification is available as a command-line tool to encode TopoJSON from GeoJSON (or ESRI Shapefiles) and a client side JavaScript library to decode TopoJSON back to GeoJSON again. TopoJSON is also supported by the popular OGR tool as of version 1.11 and PostGIS as of version 2.1.0.

TopoJSON Schema

Given a GIS shape near coordinates latitude 0° and longitude 0°, a simple but valid and complete topojson file containing all metadata, *Polygon*, *LineString*, *Point* elements, *arcs* and *properties* is defined as follows:

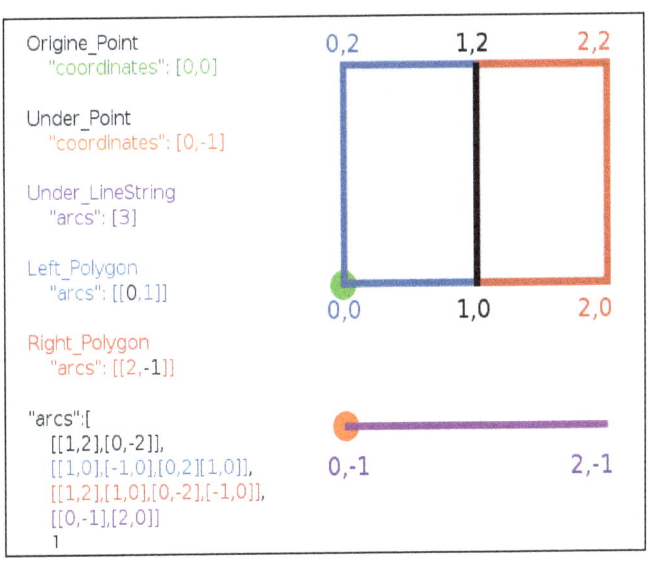

Topojson Shapes

```
{
  "type":"Topology",
  "transform":{
    "scale": [1,1],
    "translate": [0,0]
  },
  "objects":{
   "two-squares":{
     "type": "GeometryCollection",
     "geometries":[
       {"type": "Polygon", "arcs":[[0,1]],"properties": {"name": "Left_
       Polygon" }},
       {"type":   "Polygon",   "arcs":[[2,-1]],"properties":   {"name":
       "Right_Polygon" }}
   ]
  },
  "one-line": {
   "type":"GeometryCollection",
   "geometries":[
    {"type": "LineString", "arcs": ,"properties":{"name":"Under_Line-
    String"}}
   ]
  },
  "two-places":{
   "type":"GeometryCollection",
   "geometries":[
    {"type":"Point","coordinates":[0,0],"properties":{"name":"Origi-
    ne_Point"}},
    {"type":"Point","coordinates":[0,-1],"properties":{"name":"Under_
    Point"}}
```

```
      ]
    }
},
    "arcs": [
      [[1,2],[0,-2]],
      [[1,0],[-1,0],[0,2],[1,0]],
      [[1,2],[1,0],[0,-2],[-1,0]],
      [[0,-1],[2,0]]
    ]
  }
```

MapInfo TAB Format

Sometimes called the MapInfo Native format or the MapInfo TAB format, the MapInfo dataset format is a proprietary format developed for use with MapInfo software products. A MapInfo dataset consists of a number of physical files. Two files are required for textual and numeric data without graphic objects. The table structure in ASCII format (*.TAB) is the main file for a MapInfo table. The.TAB ASCII file is the link between all other files and holds information about the type of data file. The attribute data (*.DAT) is stored in binary format. These two files support a "browser" view of the tabular data, a view comparable to a spreadsheet.

File Components

The basic file components for a MapInfo Professional data set relate to the two basic environments for working in MapInfo; "Browser View" and "Mapper View".

As with most other GIS packages, several files are required to allow the user to open a data set for viewing within MapInfo Professional. The most basic view would be the browser view only. This environment provides storage of attribute or object data and is represented like a spreadsheet. In this simplified scenario, no geographic information is available.

Minimum files required for the basic MapInfo Professional browser environment:

- .TAB (The ASCII file that is the link between all other files and holds information about the type of data set file).
- .DAT (The file that stores the attribute data. This is a dBase III DBF file).

As an alternative to the *.DAT file, MapInfo Professional can use other data formats such as, *.TXT, *.XLS *.WK*, *.MDB (and for each Microsoft Access format the software also makes another small file). MapInfo Professional still creates a.TAB file that contains information about the data set file, and the user interacts with the TAB file only.

There may also be a third file:

- .IND (Optional index file for tabular data. This is present if any columns are indexed).

To view geographic information (the graphic representation of data) in MapInfo Professional, two additional files are required and added to the basic requirements for simply viewing data.

Minimum files required to view a map with the data previously discussed:

- .MAP (Stores the graphic and geographic information needed to display each vector feature on a map).
- .ID (Stores information linking graphic data to the database information. This contains a 4-byte integer index into the MAP file for each feature).

Each time a *.MAP file is saved by MapInfo Professional its objects are sorted into a storage sequence based on an R-tree spatial index. This optimises the read process of streaming data from disk to screen, at the cost of a relatively slow write process.

So the basic file set for viewing data and its graphic representation in vector form within MapInfo Professional requires a minimum of four files, normally the *.TAB, *.MAP, *.DAT and *.ID. If there is only textual information with no graphic objects, then a minimum of two files is needed: *.TAB and *.DAT.

While a data set is being edited by a user, temporary files are created by MapInfo Professional. The presence of any of these files locks the data set and prevents it from being edited by any other users. When the changes are saved, the temporary files are deleted by the software. In certain situations such as a power failure while editing is in progress, the temporary files may not be deleted and the changes will be lost. In this case the files should be manually deleted.

Normal temporary files have the following file name extensions:

- .TDA Temporary database file
- .TIN Temporary index file
- .TMA Temporary Map File

When using a database table such as Oracle Locator or Spatial, if the data is downloaded to the local machine, the temporary file name extensions are:

- .LDA Local Temporary database file
- .LIN Local Temporary index file
- .LMA Local Temporary Map File

MapInfo Professional can also use raster images and grid surfaces as layers in a map. These are stored in their native file formats, such as *.JPG, *.TIF or *.PNG, and the software again creates a *.TAB file for each one which includes georeferencing information.

A seamless table enables a single layer in a map to be derived from multiple base data sets, such as a mosaic of raster image tiles. When MapInfo Professional creates a seamless table it makes a simple vector layer which contains a rectangle for each base data set with attribute data defining the file path to the files containing the base data. When the seamless table is opened, metadata in its *.TAB file causes the map to display not the rectangles but the contents of the base tables in their place.

SOSI

SOSI is a much used geospatial vector data format for predominantly used for exchange of geographical information in Norway.

SOSI is short for Samordnet Opplegg for Stedfestet Informasjon (literally "Coordinated Approach for Spatial Information", but more commonly expanded in English to Systematic Organization of Spatial Information).

The standard includes standardized definitions for geometry and topology, data quality, coordinate systems, attributes and metadata.

The open standard was developed by the Norwegian Mapping and Cadastre Authority. It was first published in 1987 (version 1.0). It is continuously being revised and further developed. The long term development points towards international standards (ISO 19100). This work is being done by ISO/TC211, currently chaired by Olaf Østensen with the Norwegian Mapping and Cadastre Authority.

Shapefile

The ESRI Shapefile (known here as the ESRI *Shapefile format*), stores nontopological geometry and attribute information for the spatial features in a data set. A shapefile consists minimally of a main file, an index file, and a dBASE table.

In the main file, the geometry for a feature is stored as a shape comprising a set of vector coordinates. This main file is a direct access, variable-record-length file in which each record describes a shape with a list of its vertices. In the index file, each record contains the offset of the corresponding main file record from the beginning of the main file. Attributes are held in a dBASE format file. The dBASE table contains feature attributes with one record per feature. Attribute records in the dBASE file must be in the same order as records in the main file. Each attribute record has a one-to-one relationship with the associated shape record.

The shapefile format can support point, line, and area features. Area features are represented as closed loop, double-digitized polygons.

Instances of the Shapefile format have often been used as a data exchange format from ESRI formats to non-ESRI applications. The format is most useful for writing simple features and attributes quickly as there are limitations inherent in the Shapefile format related to both geometry and attributes. As outlined elsewhere in this description, these limitations may cause loss of data when using shapefiles to contain or exchange complex geometry or attributes. The Shapefile format may be used as an intermediary between data creation applications and more functionally capable GIS formats and applications, albeit with the limitations noted in the Dataset/Normal Dataset section.

The cluster of files is typically stored in the same file directory or project workspace, with all component files having the same filename (prefix) and identified by individual file extension (suffixes). Three components are mandatory: a main file that contains the feature geometry (.shp), an index file that stores the index of the feature geometry (.shx), and a dBASE table (.dbf) that stores the attribute information of features. A comprehensive list of component files follows:

- *shp* -- Main file (mandatory); a direct access, variable-record-length file in which each record describes a shape with a list of its vertices.
- *shx* -- Index file (mandatory). In the index file, each record contains the offset of the corresponding main file record from the beginning of the main file. The index file (.shx) contains a 100-byte header followed by 8-byte, fixed-length records.
- *dbf* -- dBASE Table file (mandatory); a constrained form of DBF that contains feature attributes with one record per feature. The one-to-one relationship between geometry and attributes is based on record number. Attribute records in the dBASE file must be in the same order as records in the main file.
- *sbn* -- Part 1 of spatial index for read-write instances of the Shapefile format. If present, essential for correct processing.
- *sbx* -- Part 2 of spatial index for read-write instances of the Shapefile format. If present, essential for correct processing.
- *atx* -- Created by ArcView 3.x for each instance of the Shapefile format or dBASE attribute index created in ArcCatalog. ArcView GIS 3.x attribute indexes for shapefiles and dBASE files are not used by later versions of ArcGIS as a new attribute indexing model has been developed for shapefiles and dBASE files.
- *fbn* -- One of the files that store the spatial index of the features for instances of the Shapefile format that are read-only.
- *fbx* -- The other file (besides.fbn) that stores the spatial index of the features for instances of the Shapefile format that are read-only.
- *ain* -- One of the files that stores the attribute index of the active fields in a table or a theme's attribute table.
- *aih* -- The other file (besides.ain) that stores the attribute index of the active fields in a table or a theme's attribute table.
- *ixs* -- Geocoding index for read/write shapefiles. If present, essential for correct processing.
- *mxs* --Geocoding index for read-write shapefiles (ODB format).
- *prj* -- Projections Definition file; stores coordinate system information.
- *xml* -- contains metadata, as used by ArcGIS.
- *cpg* -- An optional file that can be used to specify the codepage for identifying the character set to be used.

Shapefile Shape Format (.shp)

The main file (.shp) contains the geometry data. The binary file consists of a single fixed-length header followed by one or more variable-length records. Each of the variable-length records includes a record-header component and a record-contents component. This format should not be confused with the AutoCAD shape font source format, which shares the.shp extension.

The 2D axis ordering of coordinate data assumes a Cartesian coordinate system, using the order (X Y) or (Easting Northing). This axis order is consistent for Geographic coordinate systems, where the order is similarly (longitude latitude). Geometries may also support 3- or 4-dimensional Z and M coordinates, for elevation and measure, respectively. A Z-dimension stores the elevation of each coordinate in 3D space, which can be used for analysis or for visualisation of geometries using 3D computer graphics. The user-defined M dimension can be used for one of many functions, such as storing linear referencing measures or relative time of a feature in 4D space.

The main file header is fixed at 100 bytes in length and contains 17 fields; nine 4-byte (32-bit signed integer or int32) integer fields followed by eight 8-byte (double) signed floating point fields:

Bytes	Type	Endianness	Usage
0–3	int32	big	File code (always hex value 0x0000270a)
4–23	int32	big	Unused; five uint32
24–27	int32	big	File length (in 16-bit words, including the header)
28–31	int32	little	Version
32–35	int32	little	Shape type
36–67	double	little	Minimum bounding rectangle (MBR) of all shapes contained within the dataset; four doubles in the following order: min X, min Y, max X, max Y
68–83	double	little	Range of Z; two doubles in the following order: min Z, max Z
84–99	double	little	Range of M; two doubles in the following order: min M, max M

The file then contains any number of variable-length records. Each record is prefixed with a record header of 8 bytes:

Bytes	Type	Endianness	Usage
0–3	int32	big	Record number (1-based)
4–7	int32	big	Record length (in 16-bit words)

Following the record header is the actual record:

Bytes	Type	Endianness	Usage
0–3	int32	little	Shape type
4–	–	–	Shape content

The variable-length record contents depend on the shape type, which must be either the shape type given in the file header or Null. The following are the possible shape types:

Value	Shape type	Fields
0	Null shape	None
1	Point	X, Y
3	Polyline	MBR, Number of parts, Number of points, Parts, Points
5	Polygon	MBR, Number of parts, Number of points, Parts, Points
8	MultiPoint	MBR, Number of points, Points
11	PointZ	X, Y, Z *Optional*: M
13	PolylineZ	*Mandatory*: MBR, Number of parts, Number of points, Parts, Points, Z range, Z array *Optional*: M range, M array
15	PolygonZ	*Mandatory*: MBR, Number of parts, Number of points, Parts, Points, Z range, Z array *Optional*: M range, M array
18	MultiPointZ	*Mandatory*: MBR, Number of points, Points, Z range, Z array *Optional*: M range, M array
21	PointM	X, Y, M
23	PolylineM	*Mandatory*: MBR, Number of parts, Number of points, Parts, Points *Optional*: M range, M array
25	PolygonM	*Mandatory*: MBR, Number of parts, Number of points, Parts, Points *Optional*: M range, M array
28	MultiPointM	*Mandatory*: MBR, Number of points, Points *Optional Fields*: M range, M array
31	MultiPatch	*Mandatory*: MBR, Number of parts, Number of points, Parts, Part types, Points, Z range, Z array *Optional*: M range, M array

Shapefile Shape Index Format (.shx)

The index contains the same 100-byte header as the.shp file, followed by any number of 8-byte fixed-length records which consist of the following two fields:

Bytes	Type	Endianness	Usage
0–3	int32	big	Record offset (in 16-bit words)
4–7	int32	big	Record length (in 16-bit words)

Using this index, it is possible to seek backwards in the shapefile by, first, seeking backwards in the shape index (which is possible because it uses fixed-length records), then reading the record offset, and using that offset to seek to the correct position in the.shp file. It is also possible to seek forwards an arbitrary number of records using the same method.

Shapefile Attribute Format (.dbf)

This file stores the attributes for each shape; it uses the dBase format. An alternative format that can also be used is the xBase format, which has an open specification, and is used in open source shapefile libraries, such as the Shapefile C library.

The names and values of attributes are not standardized, and will be different depending on the source of the shapefile.

Shapefile Spatial Index Format (.sbn)

This is a binary spatial index file, which is used only by Esri software. The format is not documented by Esri. However it has been reverse-engineered and documented by the open source community. It is not currently implemented by other vendors. The.sbn file is not strictly necessary, since the.shp file contains all of the information necessary to successfully parse the spatial data.

Best Practices

Below are best practices for using shapefiles in Map Viewer.

- Choose to generalize features for web display if you have a large amount of features intended to be displayed at small scales and you have not already generalized your vertices in ArcGIS Desktop.
- Create a.zip file that contains the.shp,.shx,.dbf, and.prj files that comprise the shapefile.
- Store your shapefile directly in the root (the central directory) of the.zip archive and not in directories within the archive. If your.zip file viewer shows path information, the path should be blank.
- The name of the new layer is the same as the name of the.zip file you added. You can rename the file before adding to the map, or rename the layer once it is part of the map.
- Add standard compressed archive.zip files.
- Shapefiles that have more than 4,000 point features (or 2,000 line or polygon features) or are larger than 10 MB cannot be added directly to Map Viewer. If the file is too large, generalizing the features will often reduce the overall size and allow the shapefile to be added to the map. If that doesn't work and you have publishing privileges in your organization, you can publish your shapefile as a hosted feature layer.
- The shapefile should contain valid geometries. If you have ArcMap, you can use the Repair Geometry geoprocessing tool to correct invalid geometries in shapefiles.
- Include in the shapefile the accompanying.prj file in which the coordinate system of the data is defined. If you have ArcMap, you can define the coordinate system of a shapefile, which creates an appropriate.prj file for the shapefile.

Limitations

The following limitations apply to using shapefiles in Map Viewer:

- Compression formats other than a.zip archive are not supported.
- The shapefile must be less than 10 MB in size and have less than 4,000 point features or

2,000 line or polygon features. If you have publishing privileges in your organization, it is recommended that you publish shapefiles that exceed these limits as a hosted feature layer.

- The following features are not supported: multipatch or multipoint geometries, geometries that cross the dateline, or self-intersections in polygons. Shapefiles with these features cannot be added to a map.
- Shapefiles cannot be dragged directly onto Map Viewer; you must use the Add button in Map Viewer.

5

Analysis of GIS Database

The database that is developed to accumulate and analyze data representing objects is known as spatial database. This chapter elucidates spatial database management system, spatial data input and editing, spatial analysis and network analysis. The topics elaborated in this chapter will help in gaining a better perspective about database analysis and management.

Spatial DBMS

Computers and Information Technology have driven the human society towards a comfortable life where thousands of activities are performed on a click of a button and help save a lot of time. Everyday activities such as withdrawing money from a bank, making airline reservations, accessing a book from a library etc., are the processes that involve computer programs that access databases and database systems to facilitate all these services to the users.

The database systems have been growing and are involved in diverse applications such as:

- A multimedia database to store pictures, video clips, and sound messages.
- A real time database technology to control industrial and manufacturing processes.
- A GIS database used for storing the geographical information.
- Other common applications of databases include bank transactions, sales, reservations, registration etc.

Let us start understanding the world of databases with its functional unit known as data.

Data is defined as a collection of observations that could be in forms of facts, values, measurement, images etc. The data when used with a context becomes the information. It can be further processed, organized or simply summarized.

Given below is a list of numbers. This list of numbers is data since it communicates no meaning as such.

25
23
24
27

As soon as it is qualified that the numbers refer to the temperature recorded in degree Celsius in different cities of India on a particular day, the numbers become the temperature information.

Date	Cities	Temperature (°C)
23 October 2011	Delhi	25
	Mumbai	23
	Kolkata	24
	Chennai	27

A spatial database is a database that is optimized to store and query data that represents objects defined in a geometric space. Most spatial databases allow representing simple geometric objects such as points, lines and polygons. Some spatial databases handle more complex structures such as 3D objects, topological coverages, linear networks, and TINs. While typical databases have developed to manage various numeric and character types of data, such databases require additional functionality to process spatial data types efficiently, and developers have often added *geometry* or *feature* data types. The Open Geospatial Consortium developed the Simple Features specification (first released in 1997) and sets standards for adding spatial functionality to database systems. The *SQL/MM Spatial* ISO/EIC standard is a part the SQL/MM multimedia standard and extends the Simple Features standard with data types that support circular interpolations.

Geodatabase

A geodatabase (also geographical database and geospatial database) is a database of geographic data, such as countries, administrative divisions, cities, and related information. Such databases can be useful for websites that wish to identify the locations of their visitors for customization purposes.

Features of Spatial Databases

Database systems use indexes to quickly look up values and the way that most databases index data is not optimal for spatial queries. Instead, spatial databases use a spatial index to speed up database operations.

In addition to typical SQL queries such as SELECT statements, spatial databases can perform a wide variety of spatial operations. The following operations and many more are specified by the Open Geospatial Consortium standard:

- Spatial Measurements: Computes line length, polygon area, the distance between geometries, etc.
- Spatial Functions: Modify existing features to create new ones, for example by providing a buffer around them, intersecting features, etc.
- Spatial Predicates: Allows true/false queries about spatial relationships between geometries. Examples include "do two polygons overlap" or 'is there a residence located within a mile of the area we are planning to build the landfill?'
- Geometry Constructors: Creates new geometries, usually by specifying the vertices (points or nodes) which define the shape.
- Observer Functions: Queries which return specific information about a feature such as the location of the center of a circle.

Some databases support only simplified or modified sets of these operations, especially in cases of NoSQL systems like MongoDB and CouchDB.

Spatial Index

Spatial indices are used by spatial databases (databases which store information related to objects in space) to optimize spatial queries. Conventional index types do not efficiently handle spatial queries such as how far two points differ, or whether points fall within a spatial area of interest. Common spatial index methods include:

- HHCode
- Grid (spatial index)
- Z-order (curve)
- Quadtree
- Octree
- UB-tree
- R-tree: Typically the preferred method for indexing spatial data. Objects (shapes, lines and points) are grouped using the minimum bounding rectangle (MBR). Objects are added to an MBR within the index that will lead to the smallest increase in its size.
- R+ tree
- R* tree
- Hilbert R-tree

- X-tree
- kd-tree
- m-tree – an m-tree index can be used for the efficient resolution of similarity queries on complex objects as compared using an arbitrary metric.
- Point access method.
- Binary space partitioning (BSP-Tree): Subdividing space by hyperplanes.

Spatial Database Systems

- All OpenGIS specifications compliant products.
- Open-source spatial databases and APIs, some of which are OpenGIS-compliant.
- Caliper extends the Raima Data Manager with spatial datatypes, functions, and utilities.
- Boeing's Spatial Query Server spatially enables Sybase ASE.
- Smallworld VMDS, the native GE Smallworld GIS database.
- SpatiaLite extends Sqlite with spatial datatypes, functions, and utilities.
- IBM DB2 Spatial Extender can spatially-enable any edition of DB2, including the free DB2 Express-C, with support for spatial types.
- ClusterPoint offers native indexed support for distances, range matching and polygon matching, as well as aggregation.
- Oracle Spatial.
- Oracle Locator.
- Vertica Place, the geo-spatial extension for HP Vertica, adds OGC-compliant spatial features to the relational column-store database.
- Microsoft SQL Server has support for spatial types since version 2008.
- PostgreSQL DBMS (database management system) uses the spatial extension PostGIS to implement the standardized datatype *geometry* and corresponding functions.
- Teradata Geospatial includes 2D spatial functionality (OGC-compliant) in its data warehouse system.
- MonetDB/GIS extension for MonetDB adds OGS Simple Features to the relational column-store database.
- Linter SQL Server supports spatial types and spatial functions according to the

OpenGIS specifications.

- MySQL DBMS implements the datatype *geometry*, plus some spatial functions implemented according to the OpenGIS specifications. However, in MySQL version 5.5 and earlier, functions that test spatial relationships are limited to working with minimum bounding rectangles rather than the actual geometries. MySQL versions earlier than 5.0.16 only supported spatial data in MyISAM tables. As of MySQL 5.0.16, InnoDB, NDB, BDB, and ARCHIVE also support spatial features.
- Neo4j – a graph database that can build 1D and 2D indexes as B-tree, Quadtree and Hilbert curve directly in the graph.
- AllegroGraph – a graph database which provides a novel mechanism for efficient storage and retrieval of two-dimensional geospatial coordinates for Resource Description Framework data. It includes an extension syntax for SPARQL queries.
- MarkLogic, MongoDB, RavenDB, and RethinkDB support geospatial indexes in 2D.
- Esri has a number of both single-user and multiuser geodatabases.
- SpaceBase, a real-time spatial database.
- CouchDB a document-based database system that can be spatially enabled by a plugin called Geocouch.
- CartoDB, a cloud-based geospatial database on top of PostgreSQL with PostGIS.
- StormDB, an upcoming cloud-based database on top of PostgreSQL with geospatial capabilities.
- AsterixDB, an open-source big data management system with native geospatial capabilities.
- Kinetica, a GPU-accelerated analytics database optimized for geospatial analytics on large datasets.
- SpatialDB by MineRP, the world's first open-standards (OGC) spatial database with spatial type extensions for the Mining Industry.
- H2 supports geometry types and spatial indices as of version 1.3.173 (2013-07-28). An extension called H2GIS available on Maven Central gives full OGC Simple Features support.
- GeoMesa is a cloud-based spatio-temporal database built on top of Apache Accumulo and Apache Hadoop. GeoMesa supports full OGC Simple Features support and a GeoServer plugin.
- Ingres 10S and 10.2 include native comprehensive spatial support. Ingres in-

cludes the Geospatial Data Abstraction Library cross-platform spatial data translator.

- Tarantool supports geospatial queries with RTREE index.
- SAP HANA supports geospatial with SPS08 .
- Redis with the Geo API.

Data Storage

There are two approaches of storing data:

1. File based
2. Database

File Based Approach

File-based system is a collection of application programs that perform services for the users with each program defining and managing its data.

A flat file is an ordinary file where records of the file do not contain any information to communicate the file structure or relationship among the records to the application which is using the file.

Types of File Structure

- Unordered files
- Ordered files
- Index files

Unordered Files

	Name	Roll-ID	Class	Address
Block1	Ajay Kumar			
	Biren Das			
	Dipak Raj			
Block2	Rina Sharma			
	Prabhat Vij			
	Nira Thakur			
Block3	Tashina Rai			
	Priya Kiran			
New record	Roshan Lal			

Also known as heap files, the unordered sequential files have the basic type of organization where records are placed in the file in the order in which they are inserted i.e. new records are inserted at the end of the file.

Ordered Files

The records in such a file can be ordered based on the values of one of their fields known as ordering fields. If the ordering field is the field whose value are distinct for each individual entity of the file, then the field is known as ordering key. Reading the records in order of ordering key values becomes efficient as no sorting is required.

	Name	Roll-ID	Class	Address
Block1	Ajay Kumar			
	Biren Das			
	Dipak Raj			
Block2	Rina Sharma			
	Prabhat Vij			
	Nira Thakur			
Block3	Tashina Rai			
	Priya Kiran			

Index Files

Indexes are additional access structures which are used to speed up the retrieval of the records in response to a search condition. These provide alternative ways of accessing the records without affecting the physical placement of records.

Data Structure Models

Data models are the conceptual models that describe the structures of databases. Structure of a database is defined by the data types, the constraints and the relationships for the description or storage of data. Following are the most often used data models:

Hierarchical Data Structure Model

It is the earliest database model that is evolved from file system where records are arranged in a hierarchy or as a tree structure. Records are connected through pointers that store the address of the related record. Each pointer establishes a parent child relationship where a parent can have more than one child but a child can only have one parent. There is no connection between the elements at the same level. To locate a particular record, you have to start at the top of the tree with a parent record and trace down the tree to the child.

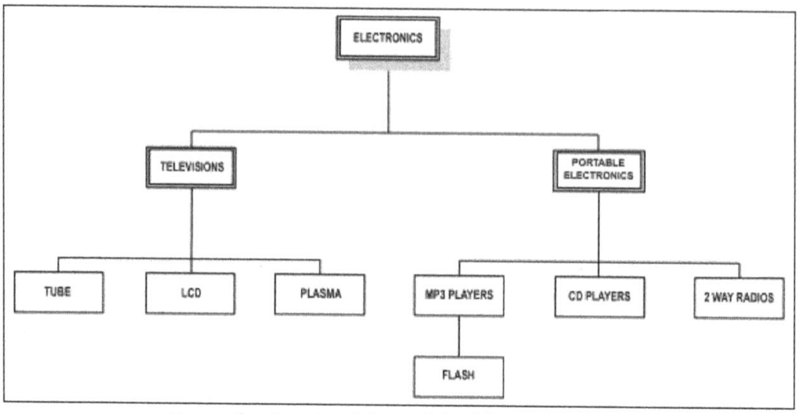

Example showing hierarchical data structure.

The figure above describes the electronic gadgets in day today use. We can see that flash is a child of mp 3 players, which is a child of portable electronics, which is a child of electronics. The topmost element electronics has no parent.

Tube, LCD, plasma, CD players and 2 way radios are leaf nodes (don't have any children).

Advantages

- Easy to understand: The organization of database parallels a family tree understanding which is quite easy.
- Accessing records or updating records are very fast since the relationships have been predefined.

Disadvantages

- Large index files are to be maintained and certain attribute values are repeated many times which lead to data redundancy and increased storage.
- The rigid structure of this model doesn't allow alteration of tables, therefore to add a new relationship entire database is to be redefined.

Network Data Structure Model

A network is a generalized graph that captures relationships between objects using connectivity. A network database consists of a collection of records that are connected to each other through links. A link is an association between two records. It allows each record to have many parents and many children thus allowing a natural model of relationships between entities.

Advantages

- The many to many relationships are easily implemented in a network data model.

- Data access and flexibility in network model is better than that in hierarchical model. An application can access an owner record and the member records within a set .
- It enforces data integrity as a user must first define owner record and then the member records.
- The model eliminated redundancy but at the expense of more complicated relationships.

Disadvantages

The network model has a complex structure that requires familiarity from user's as well as programmer's end.

Relational Data Structure Model

The relational data model was introduced by Codd in 1970. The relational database relates or connects data in different files through the use of a common field. A flat file structure is used with a relational database model. In this arrangement, data is stored in different tables made up of rows and columns. The columns of a table are named by attributes. Each row in the table is called a tuple and represents a basic fact. No two rows of the same table may have identical values in all columns.

There are two crucial data integrity constraints viz. primary key and foreign key. A primary key is an attribute whose value is unique across all tuples (rows) in a relation (table). The primary key of one table appearing as an attribute of another table is known as a foreign key in that table.

Fields/columns

Primary Key

StudentID	Age	Sex	Weight
MT001	20	F	52
MT002	21	F	55
MT003	21	M	57
MT004	22	M	60
MT005	21	M	62

Tuples/rows

Foreign Key

StudentID	Blood group
MT001	B+
MT002	B+
MT003	A+
MT004	AB+
MT005	O-

Advantages

- The manager or administrator does not have to be aware of any data structure or data pointer. One can easily add, update, delete or create records using simple logic.

Disadvantages

- A few search commands in a relational database require more time to process compared with other database models.

Working with Tables

Most databases instead of keeping their data together in a single table, organize the data into multiple tables each focusing on a specific topic. A user can link these tables if required information isn't present in a single table. The records in one table can be associated with records in the other table through a common field. The temporary associations can be made by joining and relating the tables.

Join

Joining appends the fields of one table to fields of another through an attribute/field common to both the tables.

StID	Age
N002	19
N005	21
N012	20
N015	19

Join Table

StID	Rank
N002	2
N005	3
N012	4
N015	1

Target

➡

StID	Age	Rank
N002	19	2
N005	21	3
N012	20	4
N015	19	1

Relate

Relate defines a relationship between two tables. Relates are bidirectional which means both tables involved will be able to use the relate regardless of which table owns the relate. The associated data isn't appended in the table like it is in join. However one can access the related data by selecting a particular record and then going to the related tables against that record.

Database and Database Management System

A database is a collection of logically related data. It represents an aspect of a real world and is designed, built or populated with data for a specific purpose. Many databases exist for many applications, and each one of them is maintained by a collection of programs known as a database management system.

A database management system (DBMS) is a computer program that stores and manages large amounts of data. One can define, construct, edit and share the database among various users and applications.

A database is defined by the data types, structures, and data constraints that are stored in the database. Constructing the database means storing the data in the database.

Manipulating involves querying the database to retrieve specific data, updating the database, and generation of reports. Sharing is allowing multiple users and programs to access the database concurrently. Database contents are divided into two parts viz. schema and data. Schema is the structure of database. It indicates the rule which data must obey. Data on the other hand are the facts.

Imagine that we want to store the information about the students of a particular class. This information in a database would be stored together in a single container called table. The table has rows (with different students) and columns (that contain facts on the students such as studentname, age etc.) The table is named STUDENT_INFO.

STUDENT_INFO			
STUDENTNAME	*STUDENTID*	*AGE*	*SEX*
ALICE	CE01	18	F
ANDREW	CE02	17	M
DAVID	CE03	18	M
DONA	CE04	18	F

The schema would define that STUDENT_INFO has four facts/attributes viz. ' STUDENTNAME', 'STUDENTID', 'AGE', 'SEX'. To ensure that correct data is filled in all the columns of the table one can also enforce rules and constraints for data input.

Advantages of DBMS

- Controlling Redundancy

  Redundancy means storing the same data multiple times. DBMS checks redundancy and prevents duplication of efforts, saves storage space and preserves the data files from becoming inconsistent.

- Restricting Unauthorized Access

  A DBMS provides a security and authorization system, which the database administrator uses to create accounts and to specify account restriction.

- Providing Storage Structures for Efficient Query Processing

  Database systems provide capabilities for efficient execution of queries and updates. Because the database is typically stored on disk, it provides specialized data structures to speed up disk search for the desired records. Auxiliary files called indexesare used for this purpose.

- Providing Backup and Recovery

  The backup and recovery subsystem of the DBMS helps in recovering from hardware or software failures.

- Enforcing Integrity Constraints

  Most database applications have certain integrity constraints that must be held for the data. A DBMS should provide capabilities for defining and enforcing these constraints.

Along with the advantages, the DBMS usage involves overhead costs that are not incurred in conventional file processing. These overhead costs are due to:

- High initial investment in hardware, software and training.
- Overhead for providing security, concurrency control, recovery and integrity functions.

One may use regular files instead of a DBMS under the following circumstances:

- The database and applications are simple, well defined and are not expected to change.
- Multiple user access to data isn't required.

Database Architecture

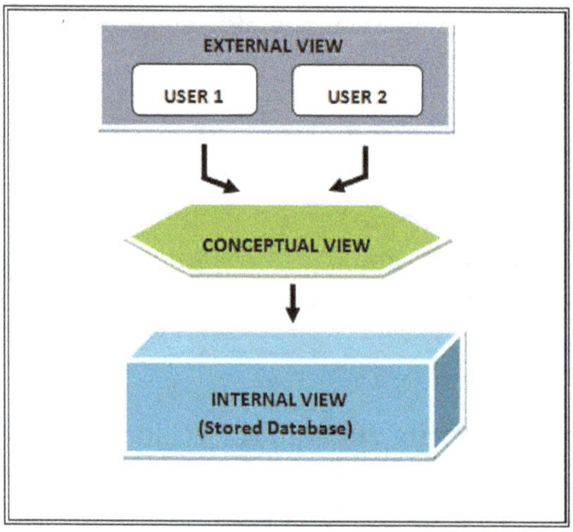

Three tier architecture of a database.

We can consider the database on three levels of abstraction:

1. *External Level* refers to user's view of the database. It describes a part of the database for particular group of users. Depending on their needs, different users access different parts of the database. It employs a powerful and flexible security mechanism by hiding parts of the database from certain users.
2. *Conceptual Level* refers to the logical structure of the entire database. It describes data as well as the relationships among the data.

3. *Internal Level* refers to the details of physical storage of the database on the computer. It consists of description of storage space allocation for data and indexes, record placements and data compression.

Normalization

Normalization is a design technique which helps designing relational databases. The objective of normalization is to create a set of relational tables that are free of redundant data and to make data consistent.

First normal form

- Eliminate repeating groups in individual tables.
- Create a separate table for each set of related data.
- Identify each set of related data with a primary key.

Second normal form

- Create separate tables for sets of values that apply to multiple records .
- Relate these tables with a foreign key.

Third normal form

- Eliminate fields that do not depend on the key.

An example explains the process of normalization which is described as following:

Unnormalized Table

Table: Student

Student#	Advisor	Adv-Room	Class1	Class2	Class3
1022	Jones	412	101-07	143-01	159-02
4123	Smith	216	201-01	211-02	214-01

The example shows a table (Table1) in which the group 'class' is mentioned three times. Since one student has several classes, these classes should be listed in a separate table. Another table in first normal form (Table2) is created by eliminating the repeating group (Class#), as shown below:

First Normalization (No Repeating Groups)

Note the Class# values for each Student# value in the above table. Class# is not functionally dependent on Student# (primary key), so this relationship is not in second normal form. Class# is separated from the first normalization table and is placed in another table (Table4).

Table: Student

Student#	Advisor	Adv-Room	Class#
1022	Jones	412	101-07
1022	Jones	412	143-01
1022	Jones	412	159-02
4123	Smith	216	201-01
4123	Smith	216	211-02
4123	Smith	216	214-01

Second Normalization (Eliminate Redundant Data)

Table: Student

Student#	Advisor	Adv-Room
1022	Jones	412
4123	Smith	216

Table: Registration

Student#	Class#
1022	101-07
1022	143-01
1022	159-02
4123	201-01
4123	211-02
4123	214-01

The attribute Adv-Room is functionally dependent on the Advisor attribute. It must be moved from the student table to some other table let's say faculty table (Table6).

Third Normalization (Eliminate Data not Dependent on Key)

Table: Student

Student#	Advisor
1022	Jones
4123	Smith

Table: Faculty

Name	Room	Dept
Jones	412	42
Smith	216	42

Object Oriented Database

According to Worboys (1995) spatial data can't be managed properly by relational data model approach. Spatial data don't naturally fit into tabular structures. More attention has been given to the development of object oriented approach of database design. The aim of object oriented model is to allow data modeling which is closer to real world. An object-oriented database uses objects as elements within database files. An object is a logical grouping of related data that represents a real world entity. Each object is a distinct entity which is identified using a key attribute called ObjectID. The object can be grouped together to form a class. Objects of the same class have same attributes, behavior and relationships with other objects.

The attributes of the object are known as its states. In addition to the states of an object information about its behavior is stored. These are the operations that are performed on the object.

For example 'hotel' is an object with states such as name, address, number of rooms. The operations that can be performed on this object can vary as plotting on map, adding to a database, increasing or decreasing number of rooms etc. The object hotel belongs to a class called 'buildings'. Any object in the class hotel will inherit the properties of the class building. Thus the states and operations applicable to 'building' will also apply to 'hotel'.

Longley et al. (2001) list the following features of object data models which make them good for modeling GIS systems:

- Encapsulation: packaging together of the description of state and behavior in each object.
- Inheritance: ability to use some or all characteristics of one object in another object.
- Polymorphism: specific implementation of operations such as create, delete etc for each object.

Spatial Data Input and Editing

Collecting data and creating a GIS database is a time consuming but an important task. There are many sources of geographic data and many ways to enter that data into a GIS. A data pool can be generated by either data capture or data transfer. The data sources are divided into following two main classes:

Primary data

It involves direct measurement of objects and phenomena. Given below is the partial list of primary data:

Remote sensing data capture: Remote sensing refers to the technique of deriving the information about the objects without getting in physical contact with them. The information is derived from the measurements of the amount of electromagnetic (EM) radiations reflected, emitted or scattered from the objects under observation. The response is measured /captured by the sensors deployed in air or in space. The remote sensing data is often talked in terms of spatial, spectral and temporal resolutions.

Spatial resolution: It refers to the size of the object that can be resolved and is the measure of the pixel size.

Spectral resolution: It refers to the wavelengths of the EM spectrum in which response of the objects is captured.

Temporal resolution: It refers to the frequency with which data is captured for the same area.

Aerial photographic data is as important as remote sensing data for a GIS project. Though both aerial photographs and remote sensing images are technically similar, they have few differences as well. The most notable difference is that aerial photographs are captured using analog optical cameras and are then rasterized by scanning a film negative. Now a days digital cameras are being used for aerial photography. The aerial photographs are suitable for surveying and mapping projects.

Both satellite images and aerial photographs can provide stereo imagery from overlapping pairs of images i.e. they can generate a three dimensional model of the earth's surface. The other advantages include global coverage and repetitive monitoring that make these datasets useful for large area projects and short time events.

Surveying: Ground surveying is based on the principle of determining the 3D location of a point with the help of angles and distance measured from other known points. Survey starts from a benchmark position. The location of all surveyed points is relative to other points. The traditional surveying involves the use of transits, theodolites, chains and tapes for angle and distance measurement. These days, electro-optical devices called total stations measure both angles as well distance to an accuracy of 1mm. Surveying is a time and resource consuming activity but is the best way of obtaining accurate geographical data.

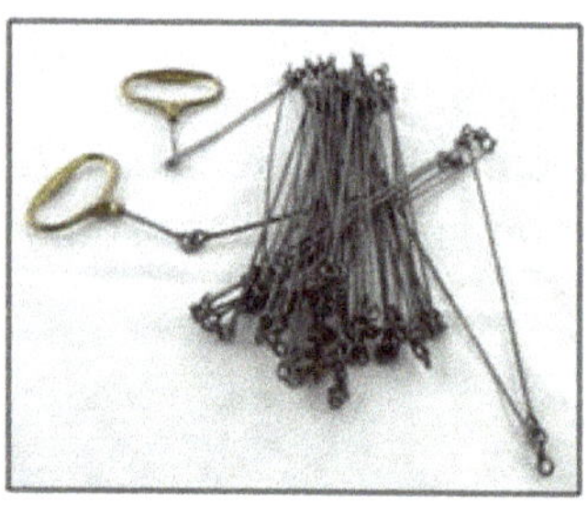

(a) Chain

(b) Theodolite

(c) Long Tape

(d) Transit Level

(e) Total station

Survey Instruments.

Sampling

Since it is not practically possible as well as worthwhile to observe the value of a variable at every point throughout the study area we adopt the strategy of sampling. Using sampling we measure subsets of the features in the area that best capture the spatial variation of the concerned attribute over the study area. The following five patterns options may be considered for sampling:

a. Simple random

This method ensures that all parts of the project area have an equal chance of being sampled. Project area is divided into a grid with numbered coordinates. A random site is picked by selecting coordinate pairs from a number table and plotting those on the project area map. Each random site is a sample point.

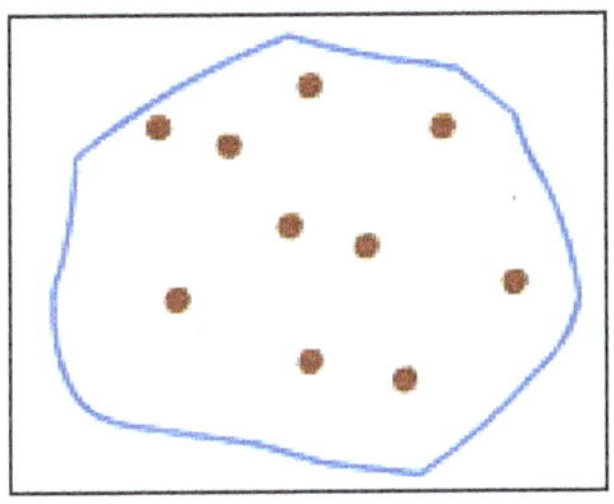

Simple random pattern.

Advantage

- Operator bias is minimal

Drawback

- Classes with small areas may be inadequately sampled or missed entirely.
- Some of the sample points may be inaccessible on ground.

b. Stratified random

It maintains randomness and at the same time overcomes the chance of an uneven distribution of points among the map classes. Specific numbers of sample points are assigned to each class with respect to its size and significance for the project. Within a class the random sites are generated in the same way as in simple random pattern.

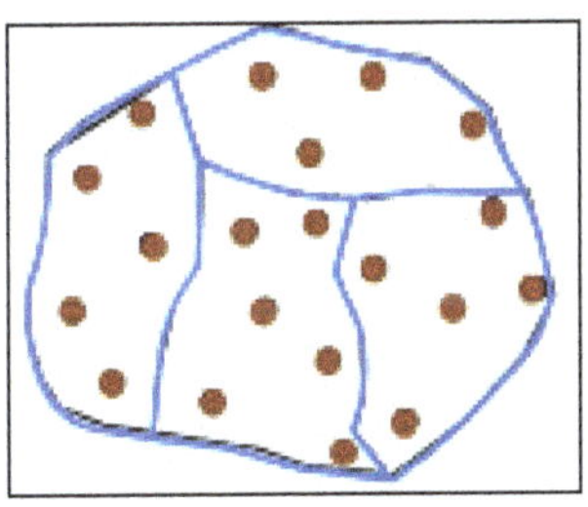

Stratified random pattern.

c. Systematic

It arranges sample points at equidistant intervals thus forming a grid. Orientation of the grid is chosen randomly.

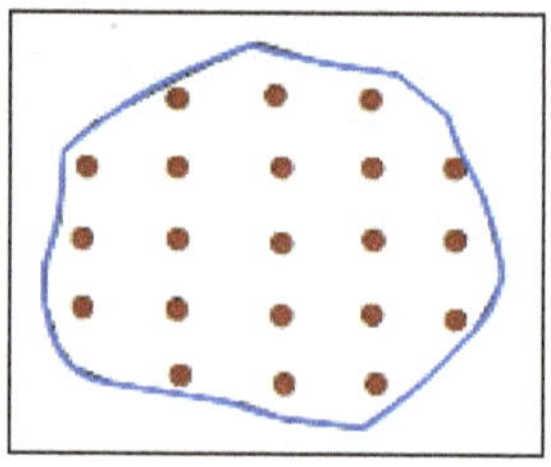

Systematic pattern.

Drawback

- Randomness is not achieved because position of every sample point is determined by the choice of the starting point.

d. Systematic unaligned

It distributes the project area into a grid and assigns the positions of sample points randomly within the grid cells.

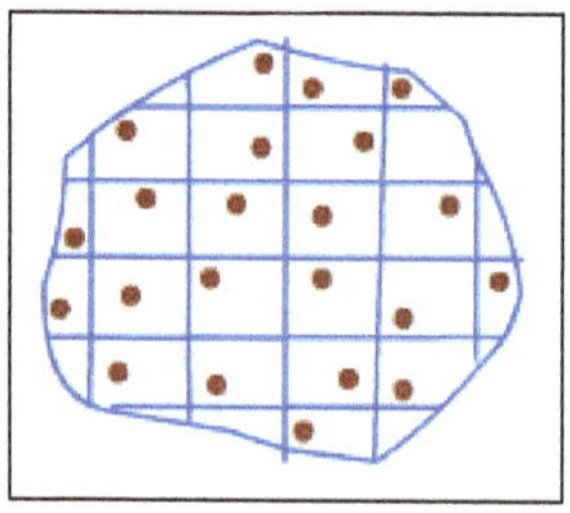

Systematic unaligned pattern.

Advantages

- All parts of the project area are sampled.
- Randomness is maintained within the grid cells.

e. Clustered

In this method, nodal points are the centers for clusters of sample points. The nodal locations are selected randomly, stratified by classes, or by identification of accessible sites.

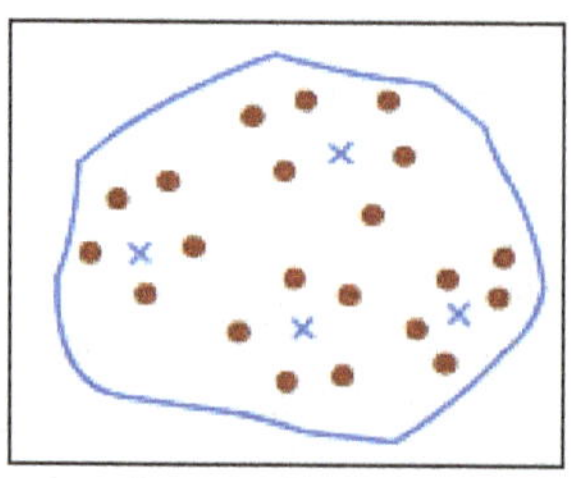

Clustered pattern.

Advantage

- In terrain with poor access, the operator can make the most of accessible sites.
- Field time is greatly reduced as lesser sites are to be visited.

GPS (Global Positioning System): GPS is a collection of 27 NAVSTAR satellites orbiting the earth at a height of 12, 500 miles or 20, 200 km. It was originally funded by US Department of Defense and was only used for military purposes but, in the year 2000 it was opened for civilians as well. GPS works on distance and time principal. The GPS satellites transmit radio signals that indicate their exact position in space. The receiver measures the time taken by signal to reach the receiver. Similarly, distance from three or more satellites helps in triangulating the position of the receiver on the earth's surface. As soon as the signal from fourth satellite is received, elevation information is also derived. GPS has led to the development of hundreds of applications affecting every aspect of modern day to day life. Farming, mining, construction, logistics, communication, power etc. are some of the sectors that have started depending heavily on GPS.

Secondary Data

It refers to the data obtained from maps, hardcopy documents etc. Some of the methods to capture secondary data are as follows:

Scanned data: A scanner is used to convert analog source map or document into digital images by scanning successive lines across a map or document and recording the amount of light reflected from the data source. Documents such as building plans, CAD drawings, images and maps are scanned prior to vectorization. Scanning helps in reducing wear and tear; improves access and provides integrated storage.

There are three different types of scanner that are widely used:

1. Flat bed scanner
2. Rotating drum scanner
3. Large format feed scanner

(a) Flat bed scanner

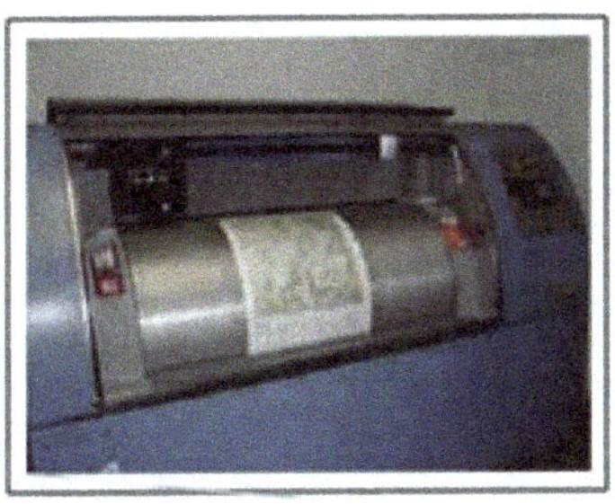

(b) Rotating drum scanner

(c) Large format feed scanner

Flat bed scanner is a PC peripheral which is small and comparatively inaccurate. The rotating drum scanners are accurate but they tend to be slow and expensive. Large format feed scanner are the most suitable type for inputting GIS data as they are cheap, quick and accurate.

Digitization: Digitizing is the process of interpreting and converting paper map or image data to vector digital data.

Heads Down Digitization

Digitizers are used to capture data from hardcopy maps. Heads down digitization is done on a digitizing table using a magnetic pen known as Puck. The position of a cursor or puck is detected when passed over a table inlaid with a fine mesh of wires. The function of a digitizer is to input correctly the coordinates of the points and the lines. Digitization can be done in two modes:

1. Point mode: In this mode, digitization is started by placing a point that marks the beginning of the feature to be digitized and after that more points are added to trace the particular feature (line or a polygon). The number of points to be added to trace the feature and the space interval between two consecutive points are decided by the operator.

2. Stream mode: In stream digitizing, the cursor is placed at the beginning of the feature, a command is then sent to the computer to place the points at either equal or unequal intervals as per the position of the cursor moving over the image of the feature.

Heads-up Digitization

This method uses scanned copy of the map or image and digitization is done on the screen of the computer monitor. The scanned map lays vertical which can be viewed without bending the head down and therefore is called as heads up digitization.

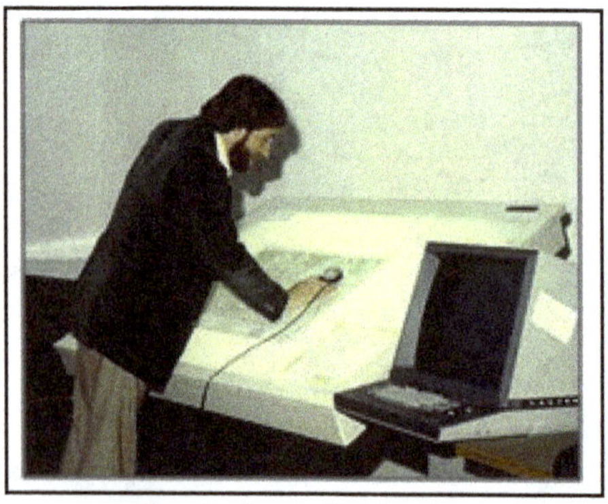

Heads down digitization.

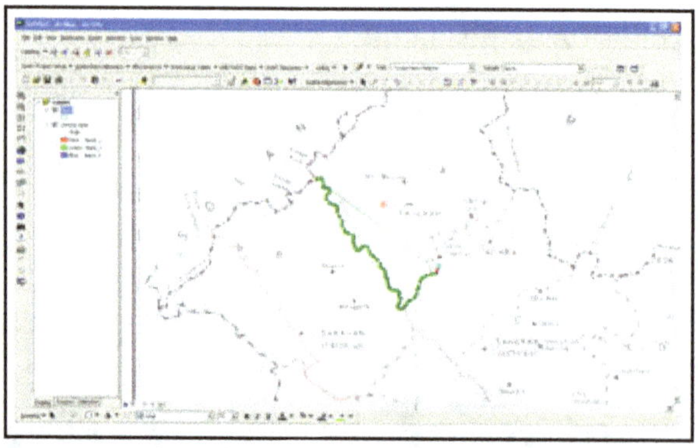

Screenshot of On-screen/Heads up digitization.

Semi-automatic and automatic methods of digitizing requires post processing but saves lot of time and resources compared to manual method and is described in the following section.

Vectorization

Vectorization is the process of converting a raster image into a vector image. It is a faster way of creating the vector data from raster data. Automatic vectorization is performed in either batch or interactive mode. Batch vectorization takes one raster file and converts it into vector objects in a single operation. Post vectorization editing is required to remove the errors. In interactive vectorization software is used to automate digitizing. The operator snaps the cursor to a pixel and indicates the direction in which line is to be digitized. The software then automatically digitizes the line. The operator can decide various parameters such as density of points, whether to pause at junction for operator's intervention or to trace in a specific direction etc. Though the process involves labor it produces high quality data and greater productivity than the manual digitization.

Photogrammetry: It is the science of making measurements from aerial photographs and images. Apart from the 2D measurement from a single photograph, photogrammetry is also used for making 3D measurements from models made using stereo pairs of photographs. To make a 3D model, there must be 60% overlap along each flight line and 30% overlap between flight lines. The measurements from overlapping pairs of photographs are captured using stereoplotters. These build a model and allow 3D measurements to be captured, edited, stored and plotted. One can extract vector objects from 3D model in a way similar to the above discussed digitization.

Obtaining Data from external sources: Creating the same dataset multiple times for the same area is a time and resource intensive process. One can always import data from data repositories. Some of these are freely available while others are available at a price. Internet is the best way to search geographic data. The internet gives information about geographic data catalogs and vendors. National agencies of a state/country also disseminate geographic data through their web portals or through other digital media on demand made by the users.

Data Editing

Errors affect the quality of GIS data. Once the data is collected, and prepared for visualization and analysis it must be checked for errors.

Burrough (1986) divided the sources of error into the following categories:

1. Common sources of error
2. Errors resulting from original measurements
3. Errors arising through processing

Common Sources of Error

- Old data sources: The data sources used for a GIS project may be too old to use. Data collected in past may not be acceptable for current time projects.
- Lack of data: The data for a given area may be incomplete or entirely lacking. For example the land-use map for border regions may not be available.
- Map scale: The details shown on a map depend on the scale used. Maps or data of the appropriate scale at which details are required, must be used for the project. Use of wrong scale would make the analysis erroneous.
- Observation: High density of observations in an area increases the reliability of the data. Insufficient observations may not provide the level of resolution required for adequate spatial analysis as expected from the project.

Errors Resulting from Original Measurements

- Positional accuracy: Representing correct positions of geographic features on map depend upon the data being used. Biased field work, improper digitization and scanning errors result in accuracies in GIS projects.
- Content accuracy: Maps must be labeled correctly. An incorrect labeling can introduce errors which may go unnoticed by the user. Any omission from map or spatial database may result in inaccurate analysis.

Errors Arising through Processing

- Numerical errors: Different computers have different capabilities for mathematical operations. Computer processing errors occur in rounding off operations and are subject to the inherent limits of number manipulation by the processor.
- Topological errors: Data is subject to variation. Errors such as dangles, slivers, overlap etc are found to be present in the GIS data layers.

Dangle: An arc is said to be a dangling arc if either it is not connected to another arc properly (undershoot) or is digitized past its intersection with another arc (overshoot).

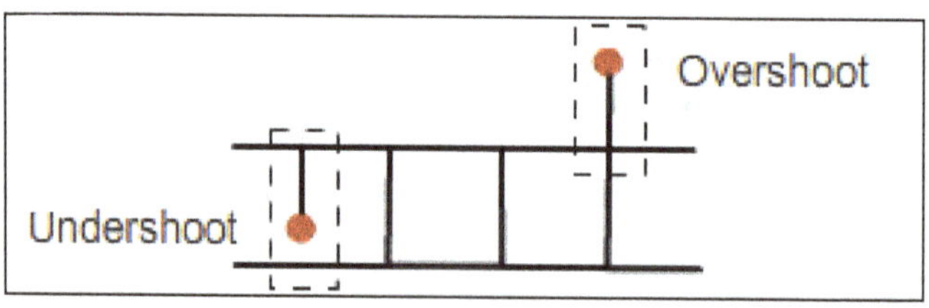

Sliver: It refers to the gap which is created between the two polygons when snapping is not considered while creating those polygons.

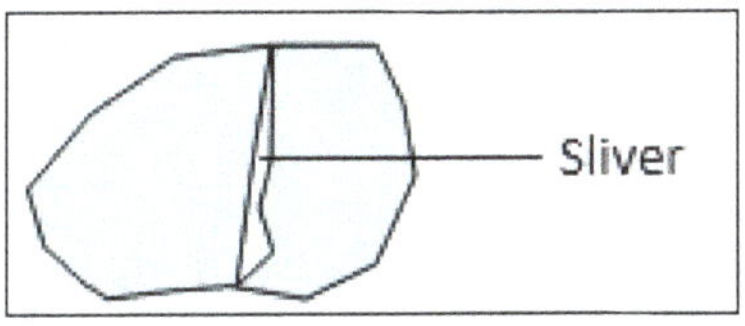

These errors can be corrected using the constraints or the rules which are defined for the layers. Topology rules define the permissible spatial relationships between features.

- Digitizing and geocoding: Many errors arise at the time of digitization, geocoding, overlaying or rasterizing. The errors associated with damaged source maps and error while digitizing can be corrected by comparing original maps with digitized versions.

Raster data editing is concerned with correcting the specific contents of raster images than their general geometric characteristics. The objective of the editing is to produce an image suitable for raster geoprocessing. Following editing functions are mostly used for raster data editing:

- Filling holes and gaps: To fill holes and gaps that appear in the raster image

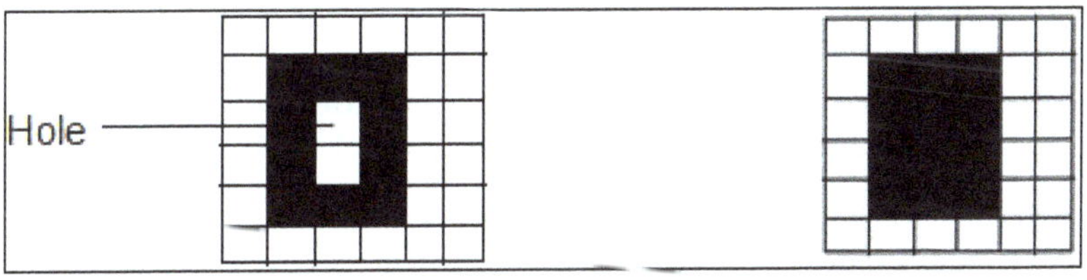

- Edge smoothing: To remove or fill single pixel irregularities in the foreground pixels and background pixels along lines

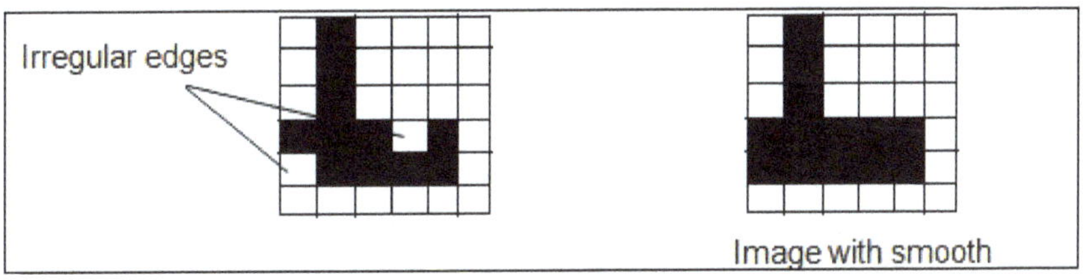

- Deskewing: To rotate the image by a small angle so that it is aligned orthogonally to the x and y axes of the computer screen

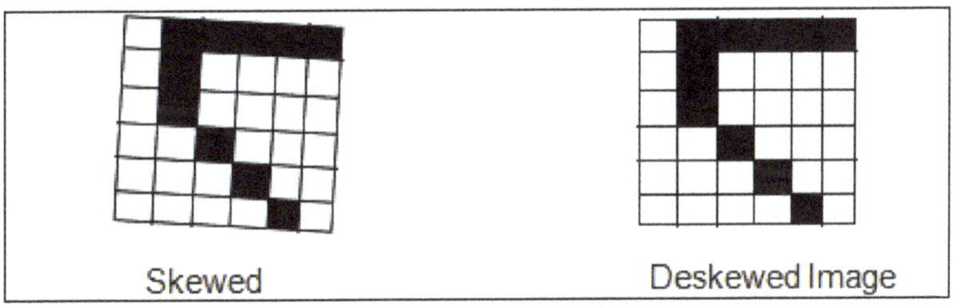

- Filtering: To remove speckles or the random high or low valued pixels in the image

- Clipping and delete: To create a subset of an image or to remove unwanted pixels

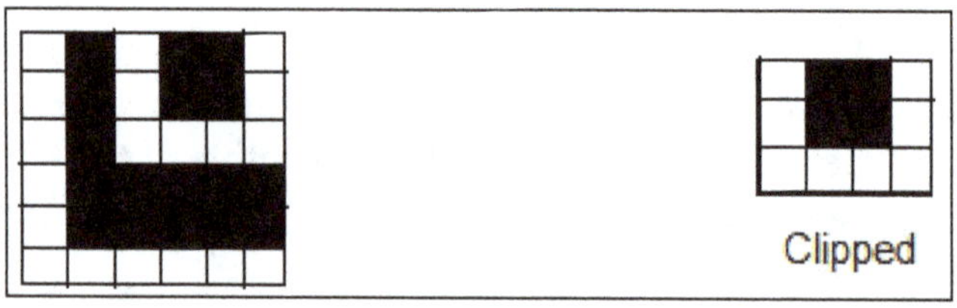

Vector data editing is a post digitizing process that ensures that the data is free from errors. It suggests that

- Lines intersect properly without having any undershoots or overshoots

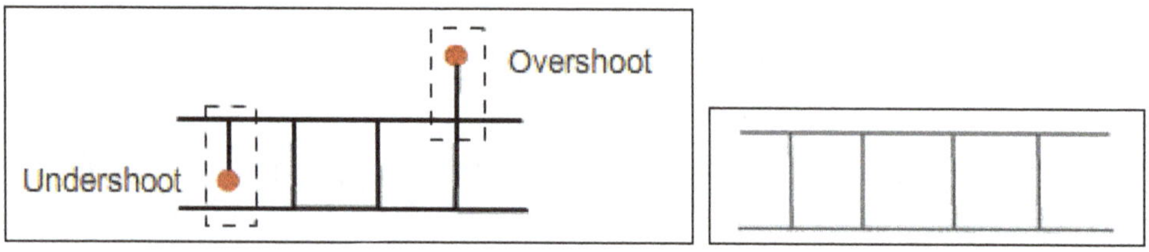

- Nodes are created at all points where lines intersect

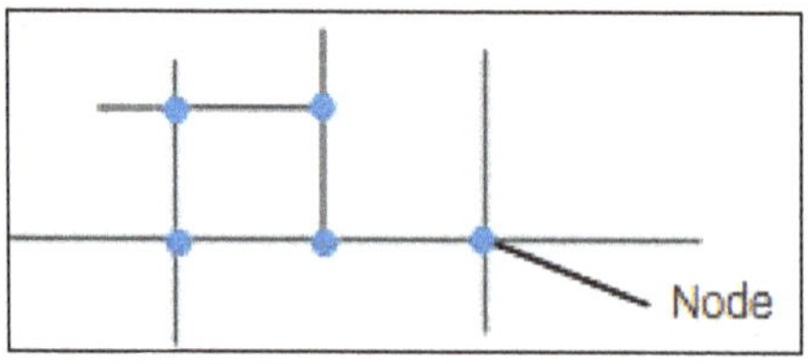

- All polygons are closed and each of them contain a label point

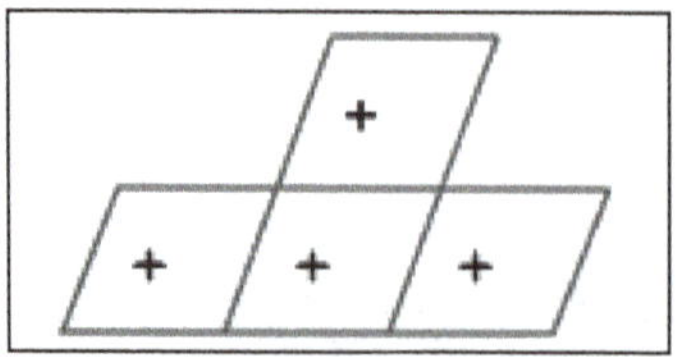

- Topology of the layer is built

Spatial Analysis

Spatial analysis is core to the GIS since all the hard work done in compiling and processing of the spatial data is to achieve this important outcome. It is only through spatial analysis that answers to a range of questions are found. Thus the spatial analysis helps in decision making.

The range of methods deployed for spatial analysis varies with respect to the type of the data model used. Before discussing these specific procedures pertaining to specific data models there are some issues that are generic and are discussed below.

Measurement: Measurement of length, perimeter and area of the features is a very common requirement in spatial analysis. However different methods are used to make measurements based on the type of data used i.e. vector or raster. Invariably, the measurements will not be exact, as digitized feature on map may not be entirely similar to the features on the ground, and moreover in the case of raster, the features are approximated using a grid cell representation.

In raster data model, the distance between two points can be calculated in the following ways:

a. Euclidean distance method: A straight line is drawn joining the two points and a right angled triangle is created. The distance is then derived using the Pythagorean geometry.

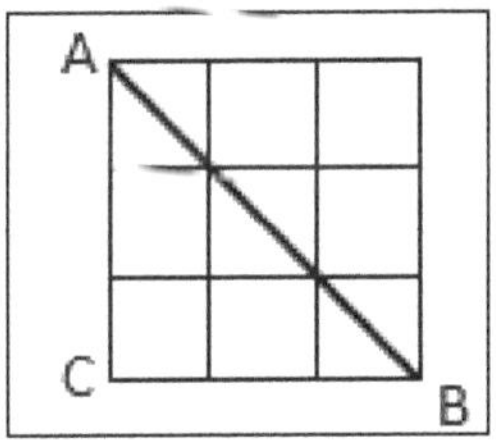

$$AB = \sqrt{AC^2 + CB^2}$$

If the cell size of the raster is 10m, then distance $AB = \sqrt{30^2 + 30^2}$

$$= \sqrt{1800}$$

$$= 42.42\text{m}$$

b. Manhattan distance method: In this method, distance along the raster cell sides from one point to the other is taken. The following example illustrates the method.

If the cell size of the raster is 10m, then distance AB = (Aa+ab+bc+cd+de+eB)

$$= (10+10+10+10+10+10)$$

$$= 60\text{m}$$

Perimeter is calculated by counting the number of cells in each side that is making the boundary of the feature and then multiplying the count by resolution (cell size) of the raster grid. All the sides of the feature are then added. Area is calculated by counting the number of cells making a feature and multiplying the count by the area of an individual grid.

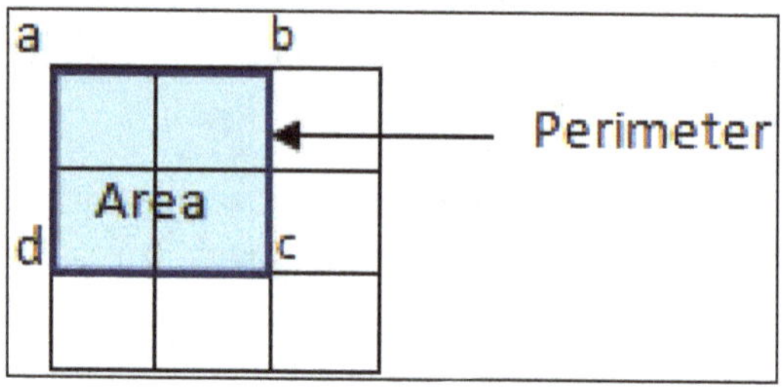

If the cell size of the raster is 10 m then the perimeter, P of the colored portion (abcd)

$$P = ab + bc + cd + ad$$
$$= (10+10)+(10+10)+(10+10)+(10+10)$$
$$= 80m$$

Since abcd is made up of 4 cells each having an area of $(10 \times 10m^2)$, therefore its area is

$$A = 4 \times 10 \times 10$$
$$= 400m^2$$

In vector data model, distance between two points is measured using the Euclidean distance method. Perimeter is calculated by adding the measurements of straight lines forming the feature. To calculate the area, the feature is subdivided into geometric shapes and then the areas of the geometric shapes are totaled.

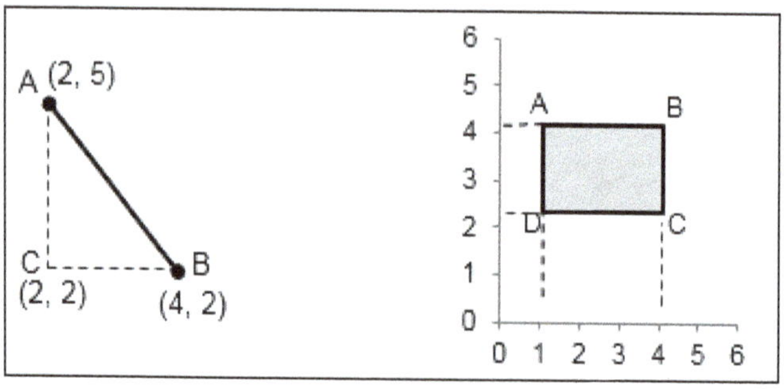

$$\text{Area of ABCD} = AB \times AD$$
$$= 3 \times 2$$
$$= 6m^2$$

$$AB = \sqrt{AC^2 + CB^2}$$
$$= \sqrt{9+4}$$
$$= 3.60m$$

$$\text{Perimeter of ABCD} = AB + BC + CD + AD$$
$$= 3 + 2 + 3 + 2$$
$$= 10m$$

Query

Query is a logical question which is performed on the database to retrieve specific data. Queries are useful for checking the quality of the data and the results obtained. There are two types of queries that can be performed in GIS:

- Aspatial or attribute queries: questions about the attributes of the feature. These do not include any spatial information. "Who owns the Star coffee shop?" is a simple query that does not involve analysis of any spatial component. Such queries could be performed by database software alone.
- Spatial queries: It involves selection of features based on location or other spatial information.

Where do the coffee shops with the same name lie in the city? Since the question asks for the location of coffee shops, the GIS software is able to show their locations on the digital map of the city.

Two or more queries can be combined together to identify features of interest. Boolean operators such as AND, OR, NOT, and XOR are used to combine queries. The spatial operations can differ depending on the data model used. The spatial operations pertaining to the vector and raster models are described below.

Map by Dr. John Snow of London, showing clusters of cholera cases in the 1854 Broad Street cholera outbreak. This was one of the first uses of map-based spatial analysis.

Spatial analysis or spatial statistics includes any of the formal techniques which study entities using their topological, geometric, or geographic properties. Spatial analysis includes a variety of techniques, many still in their early development, using different analytic approaches and applied in fields as diverse as astronomy, with its studies of the placement of galaxies in the cosmos, to chip fabrication engineering, with its use of "place and route" algorithms to build complex wiring structures. In a more restricted sense, spatial analysis is the technique applied to structures at the human scale, most notably in the analysis of geographic data.

Complex issues arise in spatial analysis, many of which are neither clearly defined nor completely resolved, but form the basis for current research. The most fundamental of these is the problem of defining the spatial location of the entities being studied.

Classification of the techniques of spatial analysis is difficult because of the large number of different fields of research involved, the different fundamental approaches which can be chosen, and the many forms the data can take.

History of Spatial Analysis

Spatial analysis can perhaps be considered to have arisen with early attempts at cartography and surveying but many fields have contributed to its rise in modern form. Biology contributed through botanical studies of global plant distributions and local plant locations, ethological studies of animal movement, landscape ecological studies of vegetation blocks, ecological studies of spatial population dynamics, and the study of biogeography. Epidemiology contributed with early work on disease mapping, notably John Snow's work of mapping an outbreak of cholera, with research on mapping the spread of disease and with location studies for health care delivery. Statistics has contributed greatly through work in spatial statistics. Economics has contributed notably through spatial econometrics. Geographic information system is currently a major contributor due to the importance of geographic software in the modern analytic toolbox. Remote sensing has contributed extensively in morphometric and clustering analysis. Computer science has contributed extensively through the study of algorithms, notably in computational geometry. Mathematics continues to provide the fundamental tools for analysis and to reveal the complexity of the spatial realm, for example, with recent work on fractals and scale invariance. Scientific modelling provides a useful framework for new approaches.

Fundamental Issues in Spatial Analysis

Spatial analysis confronts many fundamental issues in the definition of its objects of study, in the construction of the analytic operations to be used, in the use of computers for analysis, in the limitations and particularities of the analyses which are known, and in the presentation of analytic results. Many of these issues are active subjects of modern research.

Common errors often arise in spatial analysis, some due to the mathematics of space, some due to the particular ways data are presented spatially, some due to the tools which are available. Census data, because it protects individual privacy by aggregating data into local units, raises a number of statistical issues. The fractal nature of coastline makes precise measurements of its length difficult if not impossible. A computer software fitting straight lines to the curve of a coastline, can easily calculate the lengths of the lines which it defines. However these straight lines may have no inherent meaning in the real world, as was shown for the coastline of Britain.

These problems represent a challenge in spatial analysis because of the power of maps as media of presentation. When results are presented as maps, the presentation combines spatial data which are generally accurate with analytic results which may be inaccurate, leading to an impression that analytic results are more accurate than the data would indicate.

Spatial Characterization

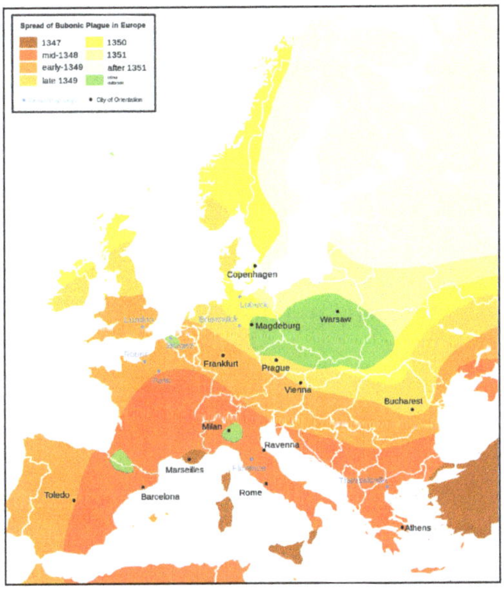

Spread of bubonic plague in medieval Europe. The colors indicate the spatial distribution of plague outbreaks over time.

The definition of the spatial presence of an entity constrains the possible analysis which can be applied to that entity and influences the final conclusions that can be reached. While this property is fundamentally true of all analysis, it is particularly important in spatial analysis because the tools to define and study entities favor specific characterizations of the entities being studied. Statistical techniques favor the spatial definition of objects as points because there are very few statistical techniques which operate directly on line, area, or volume elements. Computer tools favor the spatial definition of objects as homogeneous and separate elements because of the limited number of database elements and computational structures available, and the ease with which these primitive structures can be created.

Spatial Dependency or Auto-correlation

Spatial dependency is the co-variation of properties within geographic space: characteristics at proximal locations appear to be correlated, either positively or negatively. Spatial dependency leads to the spatial autocorrelation problem in statistics since, like temporal autocorrelation, this violates standard statistical techniques that assume independence among observations. For example, regression analyses that do not compensate for spatial dependency can have unstable parameter estimates and yield unreliable significance tests. Spatial regression models capture these relationships and do not suffer from these weaknesses. It is also appropriate to view spatial dependency as a source of information rather than something to be corrected.

Locational effects also manifest as spatial heterogeneity, or the apparent variation in a process with respect to location in geographic space. Unless a space is uniform and boundless, every location will have some degree of uniqueness relative to the other locations. This affects the spatial dependency relations and therefore the spatial process. Spatial heterogeneity means that overall parameters estimated for the entire system may not adequately describe the process at any given location.

Scaling

Spatial measurement scale is a persistent issue in spatial analysis; more detail is available at the modifiable areal unit problem (MAUP) topic entry. Landscape ecologists developed a series of scale invariant metrics for aspects of ecology that are fractal in nature. In more general terms, no scale independent method of analysis is widely agreed upon for spatial statistics.

Sampling

Spatial sampling involves determining a limited number of locations in geographic space for faithfully measuring phenomena that are subject to dependency and heterogeneity. Dependency suggests that since one location can predict the value of another location, we do not need observations in both places. But heterogeneity suggests that this relation can change across space, and therefore we cannot trust an observed degree of dependency beyond a region that may be small. Basic spatial sampling schemes include random, clustered and systematic. These basic schemes can be applied at multiple levels in a designated spatial hierarchy (e.g., urban area, city, neighborhood). It is also possible to exploit ancillary data, for example, using property values as a guide in a spatial sampling scheme to measure educational attainment and income. Spatial models such as autocorrelation statistics, regression and interpolation can also dictate sample design.

Common Errors in Spatial Analysis

The fundamental issues in spatial analysis lead to numerous problems in analysis including bias, distortion and outright errors in the conclusions reached. These issues are often interlinked but various attempts have been made to separate out particular issues from each other.

Length

In a paper by Benoit Mandelbrot on the coastline of Britain it was shown that it is inherently nonsensical to discuss certain spatial concepts despite an inherent presumption of the validity of the concept. Lengths in ecology depend directly on the scale at which they are measured and experienced. So while surveyors commonly measure the length of a river, this length only has meaning in the context of the relevance of the measuring technique to the question under study.

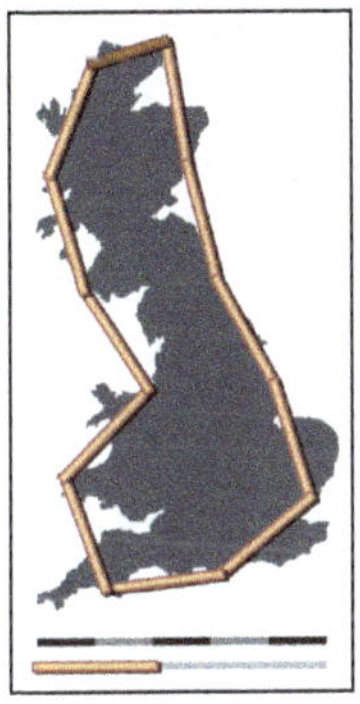

- Britain measured using a long yardstick

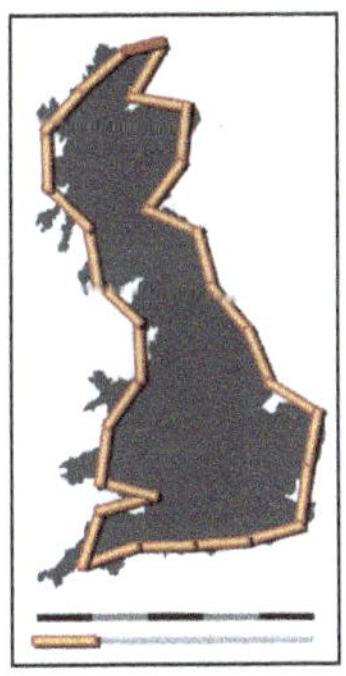

- Britain measured using a medium yardstick

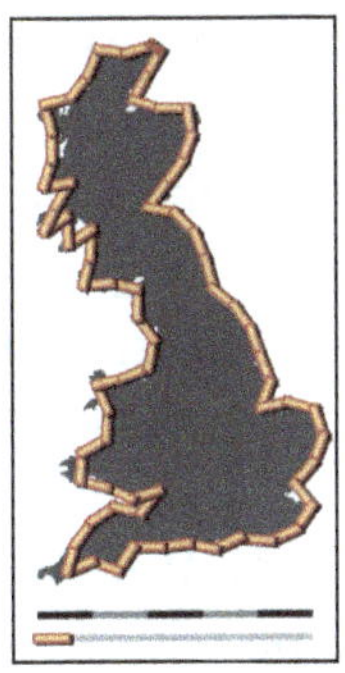

- Britain measured using a short yardstick

Locational Fallacy

The locational fallacy refers to error due to the particular spatial characterization chosen for the elements of study, in particular choice of placement for the spatial presence of the element.

Spatial characterizations may be simplistic or even wrong. Studies of humans often reduce the spatial existence of humans to a single point, for instance their home address. This can easily lead to poor analysis, for example, when considering disease transmission which can happen at work or at school and therefore far from the home.

The spatial characterization may implicitly limit the subject of study. For example, the spatial analysis of crime data has recently become popular but these studies can only describe the particular kinds of crime which can be described spatially. This leads to many maps of assault but not to any maps of embezzlement with political consequences in the conceptualization of crime and the design of policies to address the issue.

Atomic Fallacy

This describes errors due to treating elements as separate 'atoms' outside of their spatial context. The fallacy is about transferring individual conclusions to spatial units.

Ecological Fallacy

The ecological fallacy describes errors due to performing analyses on aggregate data when trying to reach conclusions on the individual units. Errors occur in part from spatial aggregation. For example, a pixel represents the average surface temperatures within an area. Ecological fallacy would be to assume that all points within the area have the same temperature. This topic is closely related to the modifiable areal unit problem.

Solutions to the Fundamental Issues

Geographic Space

A mathematical space exists whenever we have a set of observations and quantitative measures of their attributes. For example, we can represent individuals' incomes or years of education within a coordinate system where the location of each individual can be specified with respect to both dimensions. The distances between individuals within this space is a quantitative measure of their differences with respect to income and education. However, in spatial analysis we are concerned with specific types of mathematical spaces, namely, geographic space. In geographic space, the observations correspond to locations in a spatial measurement framework that captures their proximity in the real world. The locations in a spatial measurement framework often represent locations on the surface of the Earth, but this is not strictly necessary. A spatial measurement framework can also capture proximity with respect to, say, interstellar space or within a biological entity such as a liver. The fundamental tenet is Tobler's First Law of Geography: if the interrelation between entities increases with proximity in the real world, then representation in geographic space and assessment using spatial analysis techniques are appropriate.

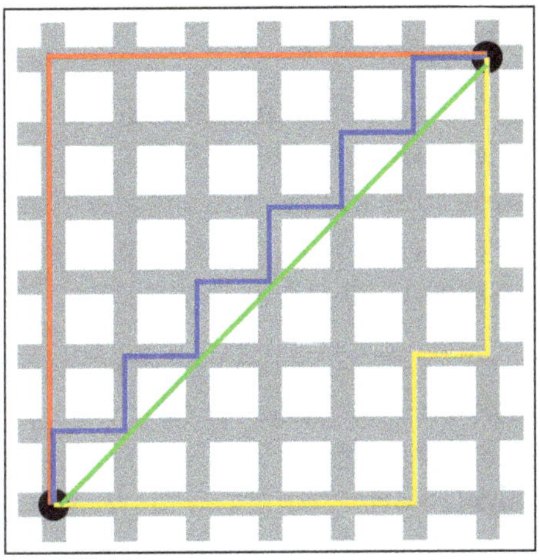

Manhattan distance versus Euclidean distance: The red, blue, and yellow lines have the same length (12) in both Euclidean and taxicab geometry. In Euclidean geometry, the green line has length $6\times\sqrt{2} \approx 8.48$, and is the unique shortest path. In taxicab geometry, the green line's length is still 12, making it no shorter than any other path shown.

The Euclidean distance between locations often represents their proximity, although this is only one possibility. There are an infinite number of distances in addition to Euclidean that can support quantitative analysis. For example, "Manhattan" (or "Taxicab") distances where movement is restricted to paths parallel to the axes can be more meaningful than Euclidean distances in urban settings. In addition to distances, other geographic relationships such as connectivity (e.g., the existence or degree of shared borders) and direction can also influence the relationships among entities. It is also possible to compute minimal cost paths across a cost surface; for example, this can represent proximity among locations when travel must occur across rugged terrain.

Types of spatial Analysis

Spatial data comes in many varieties and it is not easy to arrive at a system of classification that is simultaneously exclusive, exhaustive, imaginative, and satisfying. -- G. Upton & B. Fingelton.

Spatial Data Analysis

Urban and Regional Studies deal with large tables of spatial data obtained from censuses and surveys. It is necessary to simplify the huge amount of detailed information in order to extract the main trends. Multivariable analysis (or Factor analysis, FA) allows a change of variables, transforming the many variables of the census, usually correlated between themselves, into fewer independent "Factors" or "Principal Components" which are, actually, the eigenvectors of the data correlation matrix weighted by the inverse of their eigenvalues. This change of variables has two main advantages:

1. Since information is concentrated on the first new factors, it is possible to keep only a few of them while losing only a small amount of information; mapping them produces fewer and more significant maps.

2. The factors, actually the eigenvectors, are orthogonal by construction, i.e. not correlated. In most cases, the dominant factor (with the largest eigenvalue) is the Social Component, separating rich and poor in the city. Since factors are not-correlated, other smaller processes than social status, which would have remained hidden otherwise, appear on the second, third, ... factors.

Factor analysis depends on measuring distances between observations : the choice of a significant metric is crucial. The Euclidean metric (Principal Component Analysis), the Chi-Square distance (Correspondence Analysis) or the Generalized Mahalanobis distance (Discriminant Analysis) are among the more widely used. More complicated models, using communalities or rotations have been proposed.

Using multivariate methods in spatial analysis began really in the 1950s (although some examples go back to the beginning of the century) and culminated in the 1970s, with the increasing power and accessibility of computers. Already in 1948, in a seminal publication, two sociologists, Bell and Shevky, had shown that most city populations in the USA and in the world could be represented with three independent factors : 1- the « socio-economic status » opposing rich and poor districts and distributed in sectors running along highways from the city center, 2- the « life cycle », i.e. the age structure of households, distributed in concentric circles, and 3- « race and ethnicity », identifying patches of migrants located within the city. In 1961, in a groundbreaking study, British geographers used FA to classify British towns. Brian J Berry, at the University of Chicago, and his students made a wide use of the method, applying it to most important cities in the world and exhibiting common social structures. The use of Factor Analysis in Geography, made so easy by modern computers, has been very wide but not always very wise.

Since the vectors extracted are determined by the data matrix, it is not possible to compare factors obtained from different censuses. A solution consists in fusing together several census matrices in a unique table which, then, may be analyzed. This, however, assumes that the definition of the variables has not changed over time and produces very large tables, difficult to manage. A better solution, proposed by psychometricians, groups the data in a « cubic matrix », with three entries (for instance, locations, variables, time periods). A Three-Way Factor Analysis produces then three groups of factors related by a small cubic « core matrix ». This method, which exhibits data evolution over time, has not been widely used in geography. In Los Angeles, however, it has exhibited the role, traditionally ignored, of Downtown as an organizing center for the whole city during several decades.

Spatial Autocorrelation

Spatial autocorrelation statistics measure and analyze the degree of dependency among observations in a geographic space. Classic spatial autocorrelation statistics include Moran's I , Geary's C, Getis's G and the standard deviational ellipse. These statistics require measuring a spatial weights matrix that reflects the intensity of the geographic relationship between observations in a neighborhood, e.g., the distances between neighbors, the lengths of shared border, or whether they fall into a specified directional class such as "west". Classic spatial autocorrelation statistics compare the spatial weights to the covariance relationship at pairs of locations. Spatial autocorrelation that is more positive than expected from random indicate the clustering of similar values across geographic space, while significant negative spatial autocorrelation indicates that neighboring values are more dissimilar than expected by chance, suggesting a spatial pattern similar to a chess board.

Spatial autocorrelation statistics such as Moran's I and Geary's C are global in the sense that they estimate the overall degree of spatial autocorrelation for a dataset. The possibility of spatial heterogeneity suggests that the estimated degree of autocorrelation may vary significantly across geographic space. Local spatial autocorrelation statistics provide estimates disaggregated to the level of the spatial analysis units, allowing assessment of the dependency relationships across space. G statistics compare neighborhoods to a global average and identify local regions of strong autocorrelation. Local versions of the I and C statistics are also available.

Spatial Stratified Heterogeneity

Spatial stratified heterogeneity, referring to the within-strata variance less than the between strata-variance, is ubiquitous in ecological phenomena, such as ecological zones and many ecological variables. Spatial stratified heterogeneity reflects the essence of nature, implies potential distinct mechanisms by strata, suggests possible determinants of the observed process, allows the representativeness of observations of the earth, and enforces the applicability of statistical inferences. Spatial stratified heterogeneity of an attribute can be measured by geographical detector q-statistic:

$$q = 1 - \frac{1}{N\sigma^2} \sum_{h=1}^{L} N_h \sigma_h^2$$

where a population is partitioned into h = 1, ..., L strata; N stands for the size of the population, σ^2 stands for variance of the attribute. The value of q is within [0, 1], 0 indicates no spatial stratified heterogeneity, 1 indicates perfect spatial stratified heterogeneity. The value of q indicates the percent of the variance of an attribute explained by the stratification.The q follows a noncentral F probability density function.

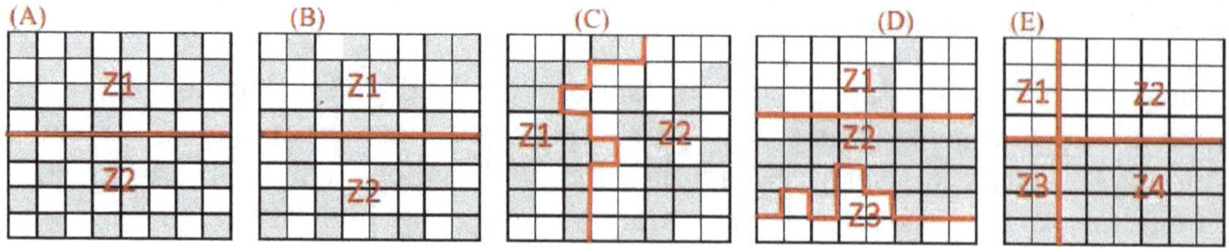

Moran's I= -1.000
q=0,p=0.9999

Moran's I= -0.393
q=0.0038,p=0.69

Moran's I= 0.000
q=0.101*, p=0.027

Moran's I= +0.393
q_1=0.16**, $p_1=3.92\text{x}10^{-3}$
q_2=0.27***, $p_2=7.01\text{x}10^{-5}$
q_3=0.67***, $p_3=4.52\text{x}10^{-15}$

Moran's I= +0.857
q_1=1***, p_1=0
q_2=0, p_2=1

A hand map with different spatial patterns. Note: p is the probability of q-statistic; * denotes statistical significant at level 0.05, ** for 0.001, *** for smaller than 10^{-3};(D) subscripts 1, 2, 3 of q and p denotes the strata Z1+Z2 with Z3,Z1 with Z2+Z3, and Z1 and Z2 and Z3 individually, respectively; (E) subscripts 1 and 2 of q and p denotes the strata Z1+Z2 with Z3+Z4,and Z1+Z3 with Z2+Z4, respectively.

Spatial Interpolation

Spatial interpolation methods estimate the variables at unobserved locations in geographic space based on the values at observed locations. Basic methods include inverse distance weighting: this attenuates the variable with decreasing proximity from the observed location. Kriging is a more sophisticated method that interpolates across space according to a spatial lag relationship that has both systematic and random components. This can accommodate a wide range of spatial relationships for the hidden values between observed locations. Kriging provides optimal estimates given the hypothesized lag relationship, and error estimates can be mapped to determine if spatial patterns exist.

Spatial Regression

Spatial regression methods capture spatial dependency in regression analysis, avoiding statistical problems such as unstable parameters and unreliable significance tests, as well as providing information on spatial relationships among the variables involved. Depending on the specific technique, spatial dependency can enter the regression model as relationships between the independent variables and the dependent, between the dependent variables and a spatial lag of itself, or in the error terms. Geographically weighted regression (GWR) is a local version of spatial regression that generates parameters disaggregated by the spatial units of analysis. This allows assessment of the spatial heterogeneity in the estimated relationships between the independent and dependent variables. The use of Bayesian hierarchical modeling in conjunction with Markov Chain Monte Carlo (MCMC) methods have recently shown to be effective in modeling complex relationships using Poisson-Gamma-CAR, Poisson-lognormal-SAR, or Overdispersed logit models. Spatial stochastic processes, such as Gaussian processes are also increasingly being deployed in spatial regression analysis. Model-based versions of GWR, known as spatially varying coefficient models have been applied to conduct Bayesian inference. Spatial stochastic process can become computationally effective and scalable Gaussian process models, such as Gaussian Predictive Processes and Nearest Neighbor Gaussian Processes (NNGP).

Spatial Interaction

Spatial interaction or "gravity models" estimate the flow of people, material or information between locations in geographic space. Factors can include origin propulsive variables such as the number of commuters in residential areas, destination attractiveness variables such as the amount of office space in employment areas, and proximity relationships between the locations measured in terms such as driving distance or travel time. In addition, the topological, or connective, relationships between areas must be identified, particularly considering the often conflicting relationship between distance and topology; for example, two spatially close neighborhoods may not display any significant interaction if they are separated by a highway. After specifying the functional forms of these relationships, the analyst can estimate model parameters using observed flow data and standard estimation techniques such as ordinary least squares or maximum likelihood. Competing destinations versions of spatial interaction models include the proximity among the destinations (or origins) in addition to the origin-destination proximity; this captures the effects of destination (origin) clustering on flows. Computational methods such as artificial neural networks can also estimate spatial interaction relationships among locations and can handle noisy and qualitative data.

Simulation and Modeling

Spatial interaction models are aggregate and top-down: they specify an overall governing relationship for flow between locations. This characteristic is also shared by urban models such as those based on mathematical programming, flows among economic sectors, or bid-rent theory. An alternative modeling perspective is to represent the system at the highest possible level of disaggregation and study the bottom-up emergence of complex patterns and relationships from behavior and interactions at the individual level.

Complex adaptive systems theory as applied to spatial analysis suggests that simple interactions among proximal entities can lead to intricate, persistent and functional spatial entities at aggregate levels. Two fundamentally spatial simulation methods are cellular automata and agent-based modeling. Cellular automata modeling imposes a fixed spatial framework such as grid cells and specifies rules that dictate the state of a cell based on the states of its neighboring cells. As time progresses, spatial patterns emerge as cells change states based on their neighbors; this alters the conditions for future time periods. For example, cells can represent locations in an urban area and their states can be different types of land use. Patterns that can emerge from the simple interactions of local land uses include office districts and urban sprawl. Agent-based modeling uses software entities (agents) that have purposeful behavior (goals) and can react, interact and modify their environment while seeking their objectives. Unlike the cells in cellular automata, simulysts can allow agents to be mobile with respect to space. For example, one could model traffic flow and dynamics using agents representing individual vehicles that try to minimize travel time between specified origins and destinations. While pursuing minimal travel times, the agents must avoid collisions with other vehicles also seeking to minimize their travel times. Cellular automata and agent-based modeling are complementary modeling strategies. They can be integrated into a common geographic automata system where some agents are fixed while others are mobile.

Multiple-point Geostatistics (MPS)

Spatial analysis of a conceptual geological model is the main purpose of any MPS algorithm. The method analyzes the spatial statistics of the geological model, called the training image, and generates realizations of the phenomena that honor those input multiple-point statistics.

A recent MPS algorithm used to accomplish this task is the pattern-based method by Honarkhah. In this method, a distance-based approach is employed to analyze the patterns in the training image. This allows the reproduction of the multiple-point statistics, and the complex geometrical features of the training image. Each output of the MPS algorithm is a realization that represents a random field. Together, several realizations may be used to quantify spatial uncertainty.

One of the recent methods is presented by Tahmasebi et al. uses a cross-correlation function to improve the spatial pattern reproduction. They call their MPS simulation method as the CCSIM algorithm. This method is able to quantify the spatial connectivity, variability and uncertainty. Furthermore, the method is not sensitive to any type of data and is able to simulate both categorical and continuous scenarios. CCSIM algorithm is able to be used for any stationary, non-stationary and multivariate systems and it can provide high quality visual appeal model.,

Geographic Information Science and spatial Analysis

Geographic information systems (GIS) and the underlying geographic information science that advances these technologies have a strong influence on spatial analysis. The increasing ability to capture and handle geographic data means that spatial analysis is occurring within increasingly data-rich environments. Geographic data capture systems include remotely sensed imagery, environmental monitoring systems such as intelligent transportation systems, and location-aware technologies such as mobile devices that can report location in near-real time. GIS provide platforms for managing these data, computing spatial relationships such as distance, connectivity and directional relationships between spatial units, and visualizing both the raw data and spatial analytic results within a cartographic context.

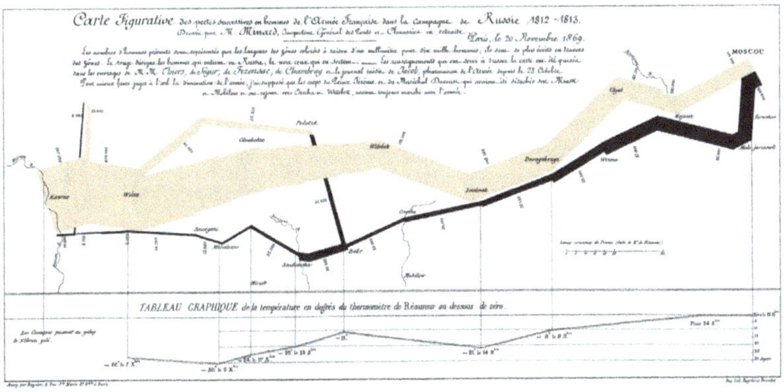

This flow map of Napoleon's ill-fated march on Moscow is an early and celebrated example of geovisualization. It shows the army's direction as it traveled, the places the troops passed through, the size of the army as troops died from hunger and wounds, and the freezing temperatures they experienced.

Geovisualization (GVis) combines scientific visualization with digital cartography to support the exploration and analysis of geographic data and information, including the results of spatial analysis or simulation. GVis leverages the human orientation towards visual information processing in the exploration, analysis and communication of geographic data and information. In contrast with traditional cartography, GVis is typically three- or four-dimensional (the latter including time) and user-interactive.

Geographic knowledge discovery (GKD) is the human-centered process of applying efficient computational tools for exploring massive spatial databases. GKD includes geographic data mining, but also encompasses related activities such as data selection, data cleaning and pre-processing, and interpretation of results. GVis can also serve a central role in the GKD process. GKD is based on the premise that massive databases contain interesting (valid, novel, useful and understandable) patterns that standard analytical techniques cannot find. GKD can serve as a hypothesis-generating process for spatial analysis, producing tentative patterns and relationships that should be confirmed using spatial analytical techniques.

Spatial decision support systems (SDSS) take existing spatial data and use a variety of mathematical models to make projections into the future. This allows urban and regional planners to test intervention decisions prior to implementation.

Vector Operations and Analysis- Single Theme

Nodepoint: Creates a new point theme from the nodes of arcs

Imagine a road theme as shown in Figure (a) having road feature- a line theme where nodes are present at locations where two or more roads meet. At these intersections lie the traffic light poles. Now if one only wants to see the location of traffic lights in the area, he/she can use the Nodepoint to extract the point theme from the nodes of the line theme to represent the location of traffic light poles as shown in Figure (b).

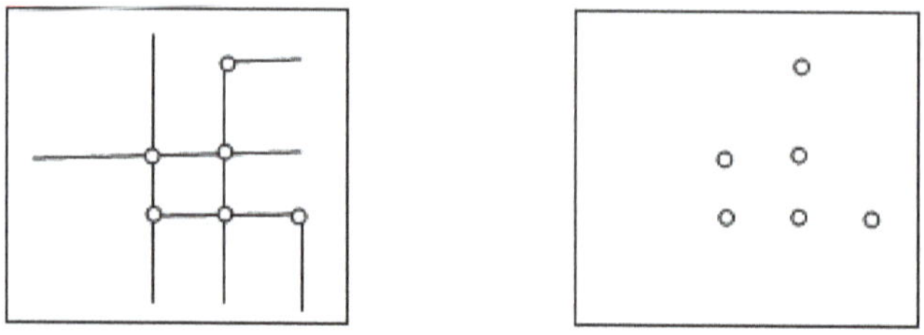

(a) (b)

Buffer: A buffer is a zone with a width created around a spatial feature and is measured in units of distance from the feature. The generated buffer takes the shape of the feature. In case of a point the buffer is a circle with a radius equal to the buffer distance. In case of a line, it is a band and for a polygon it is a belt of a specific buffer distance from the edge of polygon, surrounding the polygon. The inward buffer for a polygon is called setback (refer Figure (c), the polygon on the right hand side).

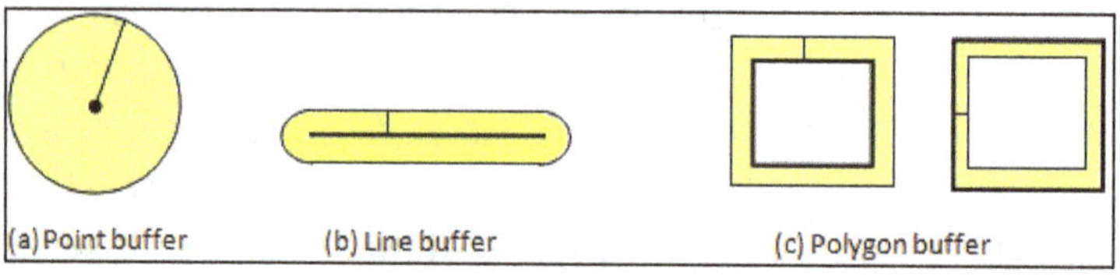

Buffering is used for neighborhood analysis which aims to evaluate the characteristics of the area surrounding the spatial feature. Common examples of buffering include the identification of properties within a certain distance of an object, delineation of areas around natural features where human activities are restricted, determination of areas affected by location etc.

Dissolve: It merges the adjacent features if they have the same attribute value and reduces the records from the attribute table. Let us understand it through following examples.

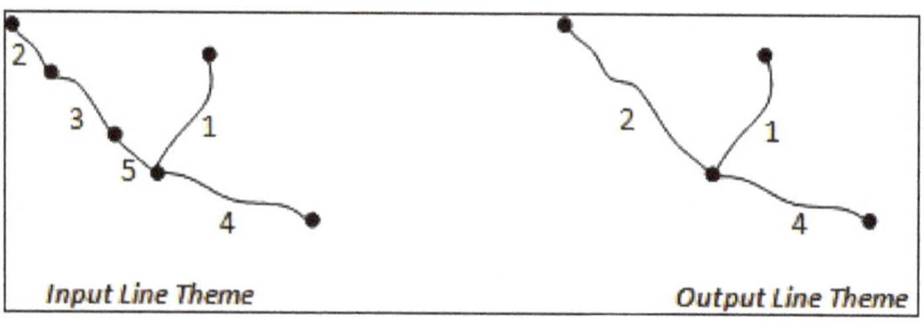

Type is used as the dissolve item.

River Attribute Table		
River#	RiverID	Type
1	1	2
2	2	1
3	3	1
4	4	2
5	5	1

New_River Attribute Table		
New_River#	New_RiverID	Type
1	1	2
2	2	1
3	4	2

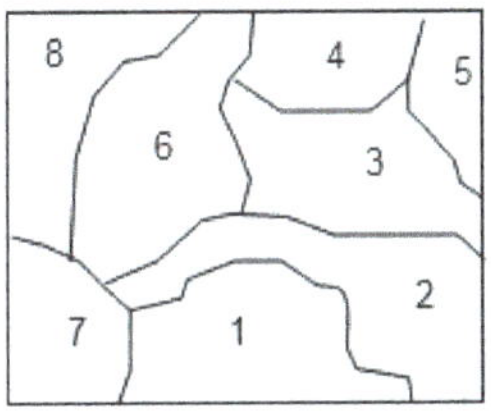

Input Polygon theme

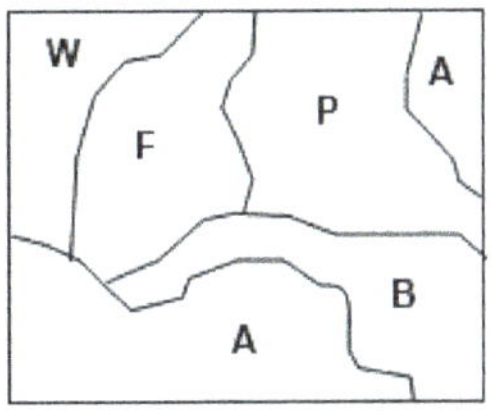

Output Polygon theme

Here, Landuse type is used as the dissolve item

Landuse Attribute table		
LUPoly#	L_ID	LanduseType
1	1	Agriculture (A)
2	2	Barren (B)
3	3	Pasture (P)
4	4	Pasture (P)
5	5	Agriculture (A)
6	6	Forest (F)
7	7	Agriculture (A)
8	8	Water (W)

New_Landuse Attribute table		
New_LUPoly#	New_LID	LanduseType
1	1	Agriculture (A)
2	2	Barren (B)
3	3	Pasture (P)
4	6	Forest (F)
5	8	Water (W)

Vector Operations and Analysis- Multiple Theme

These operations work on the layers at a time rather than selective spatial features. The layers which are used must be topologically structured so as to get a correct, topologically structured output.

Clip is used to subset a point, line or a polygon theme using another polygon theme as the boundary of the area of interest.

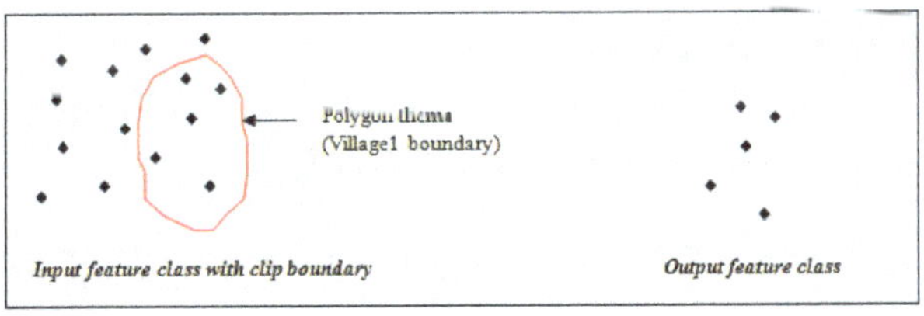

In the illustration above, the input, point feature shows the location of drinking water wells in three villages. To know how many wells fall in village1, the input feature class is clipped using the boundary of the village1. The output feature class shows that five wells are present in village1.

Split causes the input features to form subset of multiple output feature classes. The split field's unique values form the names of the output feature classes.

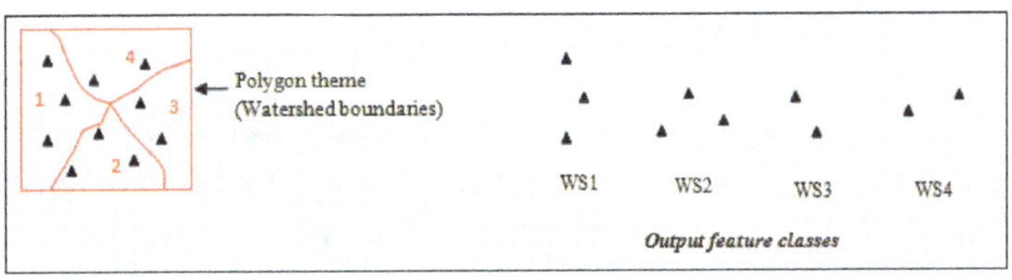

In the illustration above, a point theme of wells is split using the polygon theme of watershed boundaries. The output of this operation contains multiple feature classes which are named on the unique value of watershed boundaries (in this case, the unique value is the watershed number WS1, WS2 etc.). Each output class represents the number of wells present in a particular watershed i.e. WS1 or watershed 1 has three wells. Similarly, WS2, WS3 and WS4 have 3, 2, and 2 wells respectively.

Overlay Operations

Union creates a new theme by overlaying two polygon themes. It is same as 'or' Boolean operator. The output theme contains the combined polygons and attributes of both themes. Only polygon themes can be combined using union.

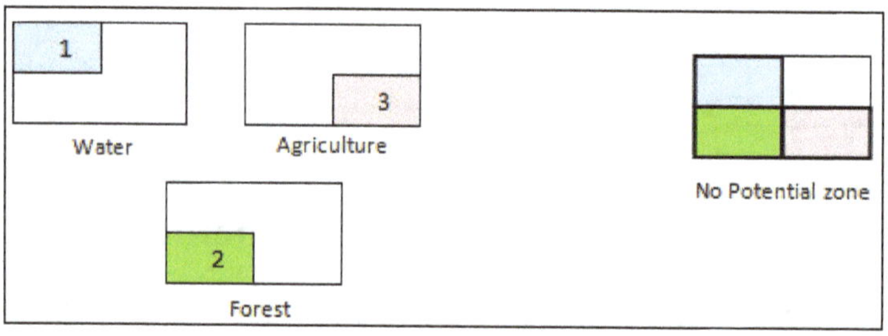

Let's say we are interested in knowing no potential zone for urban development. It is clear that no construction can be done on a waterbody or land covered by agriculture or forest. So, we can say union of areas under water, agriculture and forest would present us the area having no potential for urban development.

Intersect creates a new theme by overlaying a point, line or polygon theme with an intersecting polygon theme. It is same as 'and' Boolean operator. The output theme contains only the feature inside the intersecting polygons.

From the same example given above, if we try to know the area having potential for urban development we need to intersect the polygon themes to get a common area which is not under water, agriculture or forest.

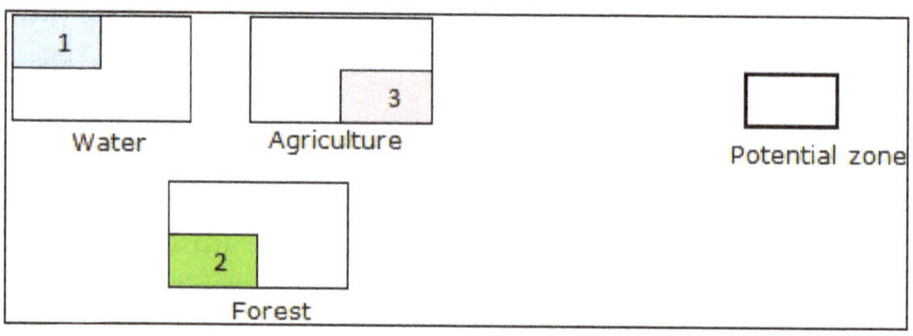

Identity creates a new theme by overlaying a point, line, or polygon theme with an intersecting polygon theme. The output theme contains all the original point, line, or polygons as well as the attributes transferred by the intersecting polygon theme.

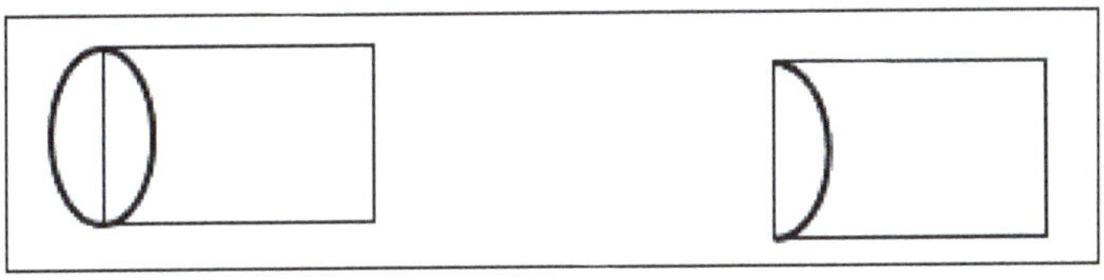

Erase removes polygons in a theme from the area covered by polygons of another theme. The output feature class only contains those features of the input polygon theme that fall outside the polygons of the second theme.

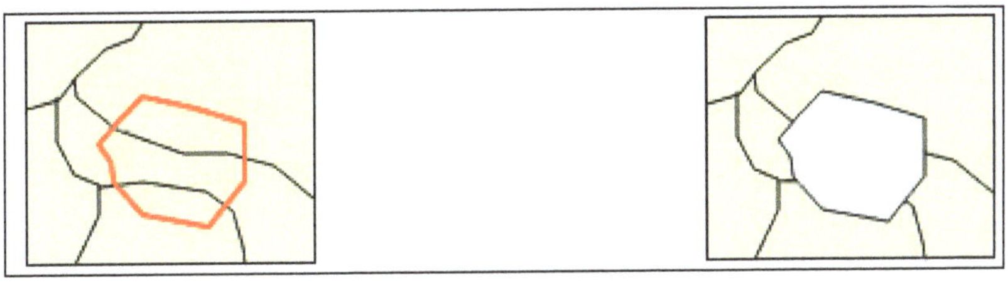

Network Analysis

It is a type of line analysis which involves set of interconnected lines. Railways, highways, transportation routes, rivers etc are examples of networks. Network analysis can be used for the following:

Address Geocoding

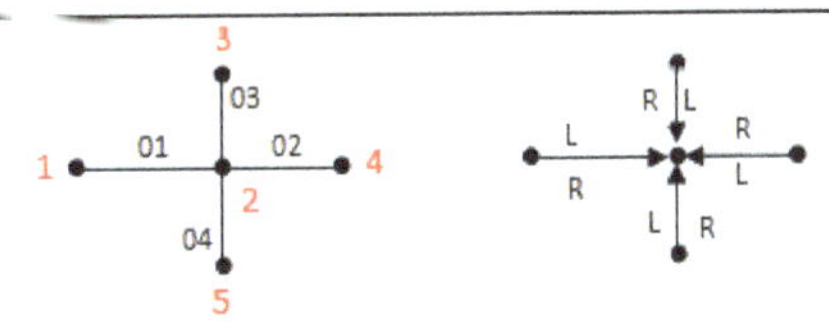

Arc#	From_node	To_node
01	1	2
02	4	2
03	3	2
04	5	2

Arc	Length	Left_from	Left_to	Right_from	Right_to	Prefix	Name	Type	Suffix
01	2000	100	1300	101	1299	West	Main	Street	
02	1800	2801	1401	2800	1400	East	Main	Street	
03	1000	200	248	201	249		First	Avenue	North
04	800	299	251	300	250		First	Avenue	South

It is the process of estimating the locations of addresses in GIS coordinate system. It requires a table of addresses and theme that contains attributes that can be used to match to the table of addresses.

Imagine that the fire department is reported about a fire in a building at 1000 West Main Street. To estimate the location, GIS determines the arc by matching its name, type and suffix. Once the arc is determined the address can be estimated using linear interpolation. The arc corresponding to West Main Street is the Arc 01.

The address is an even number and lies on the left side of the arc. The left side has addresses ranging from 100 to 1300 (range is 1200). The length of the arc is 2000 meters. The address of 1000 can be geocoded as:

$$\frac{(\text{Max. address} - \text{address})}{\text{Range}} \times \text{Arc length}$$

$$= \frac{(1300-1000)}{1200} \times 2000$$

$$= 500 \text{ metres along the arc}$$

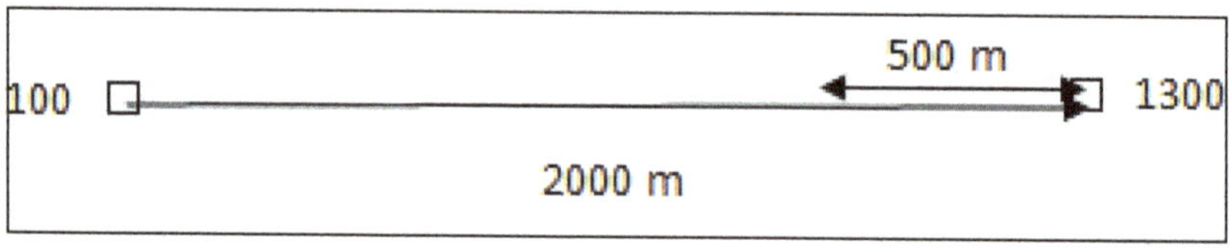

Optimal Routing

Optimal routing is the process of finding out the best route to go from one location to another location. The most common path finding algorithm is Dijkstra algorithm which was published by E.W. Dijkstra in 1956. It is a graph search algorithm that provides the shortest path for a single source shortest path problem.

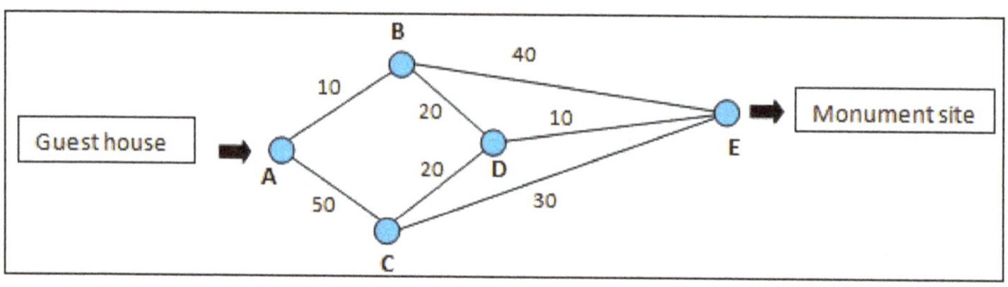

Imagine we have the following network of roads. We want to find the quickest path to get from the guest house to monument site. The variation in time travelling along different roads depends upon the traffic conditions.

We build two tables, one for the nodes that have been already processed and the other for the adjacent nodes which are to be processed. We begin with Node A as follows:

Processed Nodes		
Node	Cumulative cost	Previous node
A	0	none

Adjacent Nodes		
Node	Cumulative cost	Previous node
B	10	A
C	50	A

We pick the adjacent node with least cumulative cost, and place it in the processed node table.

Processed Nodes		
Node	Cumulative cost	Previous node
A	0	none
B	10	A

Adjacent Nodes		
Node	Cumulative cost	Previous node
C	50	A
D	30	B
E	50	B

Processed Nodes		
Node	Cumulative cost	Previous node
A	0	none
B	10	A
D	30	B

Adjacent Nodes		
Node	Cumulative cost	Previous node
C	50	A
E	50	B
C	50	D
E	40	D

Processed Nodes		
Node	Cumulative cost	Previous node
A	0	none
B	10	A
D	30	B
E	40	D

Nodc E is the destination so we stop here. The quickest route to reach node E takes 40 minutes and it is Node A → Nodc B → Node C → Node D.

Finding Closest Facilities

Sometimes we try to find out a point closest to a given location. The point is called a facility and the given location is called an event location. Finding which flat would be near to the working place, which fire station has the best response time to a report to the fire location, which houses are close to the schools are examples of optimal routing for closest facilities.

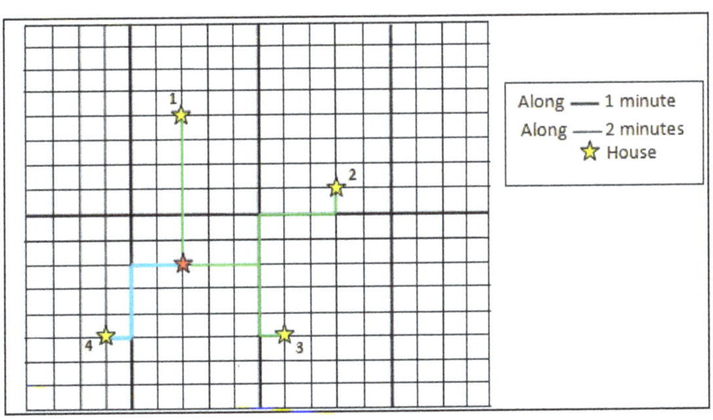

In the above illustration, imagine that the office (red star) is located at some address and we want to know the closest place around the office where an employee can live. The address of the office is geocoded to a street location and then optimal path can be computed from each house to the office.

House	Estimated Travel Time (min)
1	12
2	13
3	11
4	9

House 4 is closest to the office as the travel time is least for it.

Geocoding

Geocoding is the computational process of transforming a postal address description to a location on the Earth's surface (spatial representation in numerical coordinates). Reverse geocoding, on the other hand, converts the inputted geographic coordinates to a description of a location, usually the name of a place or a postal address. Geocoding relies on a computer representation of the street network. Geocoding is sometimes used for conversion from ZIP codes or postal codes to coordinates, occasionally for the conversion of parcel identifiers to centroid coordinates.

Geocoding (*verb*): The act of transforming an address text into a valid spatial representation.

Geocoder (*noun*): A piece of software or a (web) service that implements a geocoding process i.e. a set of interrelated components in the form of operations, algorithms, and data sources that work together to produce a spatial representation for descriptive locational references.

Geocode (*noun*): A spatial representation of a descriptive locational reference.

The geographic coordinates representing locations often vary greatly in positional accuracy. Examples include building centroids, land parcels, street addresses, postal code centroids (e.g. ZIP codes, CEDEX), and Administrative Boundary Centroids.

History

Geocoding — a subset of Geographic Information System (GIS) spatial analysis — has been a subject of interest since the early 1960s.

1960s

In 1960, the first operational GIS – named the Canada Geographic Information System (CGIS) – was invented by Dr. Roger Tomlinson, who has since been acknowledged as the father of GIS. The CGIS was used to store and analyze data collected for the Canada Land Inventory, which mapped information about agriculture, wildlife, and forestry at a scale of 1:50,000, in order to regulate land capability for rural Canada. However, the CGIS lasted until the 1990s and was never available commercially.

On July 1, 1963, five-digit ZIP codes were introduced nationwide by the United States Post Office Department (USPOD). In 1983, nine-digit ZIP+4 codes were brought about as an extra identifier in more accurately locating addresses.

In 1964, the Harvard Laboratory for Computer Graphics and Spatial Analysis developed groundbreaking software code — e.g. GRID, and SYMAP — all of which were sources for commercial development of GIS.

In 1967, a team at the Census Bureau — including the mathematician James Corbett and Donald Cooke — invented Dual Independent Map Encoding (DIME) — the first modern vector mapping model — which ciphered address ranges into street network files and incorporated the "percent along" geocoding algorithm. Still in use by platforms such as Google Maps and MapQuest, the "percent along" algorithm denotes where a matched address is located along a reference feature as a percentage of the reference feature's total length. DIME was intended for the use of the United States Census Bureau, and it involved accurately mapping block faces, digitizing nodes representing street intersections, and forming spatial relationships. New Haven, Connecticut was the first city on Earth with a geocodable streets network database.

1980s

In the late 1970s, two main public domain geocoding platforms were in development: GRASS GIS and MOSS. The early 1980s saw the rise of many more commercial vendors of geocoding software, namely Intergraph, ESRI, CARIS, ERDAS, and MapInfo Corporation. These platforms merged the 1960s approach of separating spatial information with the approach of organizing this spatial information into database structures.

In 1986, Mapping Display and Analysis System (MIDAS) became the first desktop geocoding software, designed for the DOS operating system. Geocoding was elevated from the research department into the business world with the acquisition of MIDAS by MapInfo. MapInfo has since been acquired by Pitney Bowes, and has pioneered in merging geocoding with business intelligence; allowing location intelligence to provide solutions for the public and private sectors.

1990s

The end of the 20th century had seen geocoding become more user-oriented, especially via open-source GIS software. Mapping applications and geospatial data had become more accessible over the Internet.

Because the mail-out/mail-back technique was so successful in the 1980 Census, the U.S. Bureau of Census was able to put together a large geospatial database, using interpolated street geocoding. This database — along with the Census' nationwide coverage of households — allowed for the birth of TIGER (Topologically Integrated Geographic Encoding and Referencing).

Containing address ranges instead of individual addresses, TIGER has since been implemented in nearly all geocoding software platforms used today. By the end of the 1990 Census, TIGER "contained a latitude/longitude-coordinate for more than 30 million feature intersections and endpoints and nearly 145 million feature 'shape' points that defined the more than 42 million feature segments that outlined more than 12 million polygons."

TIGER was the breakthrough for "big data" geospatial solutions.

2000s

The early 2000s saw the rise of Coding Accuracy Support System (CASS) address standardization. The CASS certification is offered to all software vendors and advertising mailers who want the United States Postal Services (USPS) to assess the quality of their address-standardizing software. The annually renewed CASS certification is based on delivery point codes, ZIP codes, and ZIP+4 codes. Adoption of a CASS certified software by software vendors allows them to receive discounts in bulk mailing and shipping costs. They can benefit from increased accuracy and efficiency in those bulk mailings, after having a certified database. In the early 2000s, geocoding platforms were also able to support multiple datasets.

In 2003, geocoding platforms were capable of merging postal codes with street data, updated monthly. This process became known as "conflation".

Beginning in 2005, geocoding platforms included parcel-centroid geocoding. Parcel-centroid geocoding allowed for a lot of precision in geocoding an address. For example, parcel-centroid allowed a geocoder to determine the centroid of a specific building or lot of land. Platforms were now also able to determine the elevation of specific parcels.

2005 also saw the introduction of the Assessor's Parcel Number (APN). A jurisdiction's tax assessor was able to assign this number to parcels of real estate. This allowed for proper identification and record-keeping. An APN is important for geocoding an area which is covered by a gas or oil lease, and indexing property tax information provided to the public.

In 2006, Reverse Geocoding and reverse APN lookup were introduced to geocoding platforms. This involved geocoding a numerical point location — with a longitude and latitude — to a textual, readable address.

2008 and 2009 saw the growth of interactive, user-oriented geocoding platforms – namely MapQuest, Google Maps, Bing Maps, and Global Positioning Systems (GPS). These platforms were made even more accessible to the public with the simultaneous growth of the mobile industry, specifically smartphones.

2010s

This current decade has seen vendors fully supporting geocoding and reverse geocoding globally. Cloud-based geocoding application programming interface (API) and on-premise geocoding has allowed for a greater match rate, greater precision, and greater speed. There is now a popularity in the idea of geocoding being able to influence business decisions. This is the integration between the geocoding process and business intelligence.

The future of geocoding also involves three-dimensional geocoding, indoor geocoding, and multiple language returns for the geocoding platforms.

Geocoding Process

Geocoding is a task which involves multiple datasets and processes, all of which work together. A geocoder is made of two important components: a reference dataset and the geocoding algorithm. Each of these components are made up of sub-operations and sub-components. Without understanding how these geocoding processes work, it is difficult to make informed business decisions based on geocoding.

Input Data

Input data are the descriptive, textual information (address or building name) which the user wants to turn into numerical, spatial data (latitude and longitude) – through the process of geocoding.

Classification of Input Data

Input data is classified into two categories: relative input data and absolute input data.

Relative Input Data

Relative input data are the textual descriptions of a location which, alone, cannot output a spatial representation of that location. Such data outputs a relative geocode, which is dependent and geographically relative of other reference locations. An example of a relative geocode is address-interpolation using areal units or line vectors. "Across the street from the Empire State Building" is an example of a relative input data. The location being sought cannot be determined without identifying the Empire State Building. Geocoding platforms often do not support such relative locations, but advances are being made in this direction.

Absolute Input Data

Absolute input data are the textual descriptions of a location which, alone, can output a spatial representation of that location. This data type outputs an absolute known location independently of other locations. For example, USPS ZIP codes; USPS ZIP+4 codes; complete and partial postal addresses; USPS PO boxes; rural routes; cities; counties; intersections; and named places can all be referenced in a data source absolutely.

When there is a lot of variability in the way addresses can be represented — such as too much input data or too little input data — geocoders use address normalization and address standardization in order to resolve this problem.

Processing of Input Data

Address interpolation

A simple method of geocoding is address interpolation. This method makes use of data from a street geographic information system where the street network is already mapped within the geographic coordinate space. Each street segment is attributed with address ranges (e.g. house numbers from one segment to the next). Geocoding takes an address, matches it to a street and specific segment (such as a block, in towns that use the "block" convention). Geocoding then interpolates the position of the address, within the range along the segment.

Example

Take for example: *742 Evergreen Terrace*

Let's say that this segment (for instance, a block) of Evergreen Terrace runs from 700 to 799. Even-numbered addresses fall on the east side of Evergreen Terrace, with odd-numbered addresses on the west side of the street. 742 Evergreen Terrace would (probably) be located slightly less than halfway up the block, on the east side of the street. A point would be mapped at that location along the street, perhaps offset a distance to the east of the street centerline.

Complicating Factors

However, this process is not always as straightforward as in this example. Difficulties arise when

- distinguishing between ambiguous addresses such as 742 Evergreen Terrace and 742 W Evergreen Terrace.
- attempting to geocode new addresses for a street that is not yet added to the geographic information system database.

While there might be 742 Evergreen Terrace in Springfield, there might also be a 742 Evergreen Terrace in Shelbyville. Asking for the city name (and state, province, country, etc. as needed) can solve this problem. Boston, Massachusetts has multiple "100 Washington Street" locations because several cities have been annexed without changing street names, thus requiring use of unique postal codes or district names for disambiguation. Geocoding accuracy can be greatly improved by first utilizing good address verification practices. Address verification will confirm the existence of the address and will eliminate ambiguities. Once the valid address is determined, it is very easy to geocode and determine the latitude/longitude coordinates. Finally, several caveats on using interpolation:

- The typical attribution of a street segment assumes that all even numbered parcels are on one side of the segment, and all odd numbered parcels are on the other. This is often not true in real life.
- Interpolation assumes that the given parcels are evenly distributed along the length of the segment. This is almost never true in real life; it is not uncommon for a geocoded address to be off by several thousand feet.
- Interpolation also assumes that the street is straight. If a street is curved then the geocoded location will not necessarily fit the physical location of the address.
- Segment Information (esp. from sources such as TIGER) includes a maximum upper bound for addresses and is interpolated as though the full address range is used. For example, a segment (block) might have a listed range of 100-199, but the last address at the end of the block is 110. In this case, address 110 would be geocoded to 10% of the distance down the segment rather than near the end.
- Most interpolation implementations will produce a point as their resulting address location. In reality, the physical address is distributed along the length of the segment, i.e. consider geocoding the address of a shopping mall - the physical lot may run a distance along the street segment (or could be thought of as a two-dimensional space-filling polygon which may front on several different streets — or worse, for cities with multi-level streets, a three-dimensional shape that meets different streets at several different levels) but the interpolation treats it as a singularity.

A very common error is to believe the accuracy ratings of a given map's geocodable attributes. Such accuracy currently touted by most vendors has no bearing on an address being attributed to the correct segment, being attributed to the correct side of the segment, nor resulting in an accurate position along that correct segment. With the geocoding process used for U.S. Census TIGER datasets, 5-7.5% of the addresses may be allocated to a different census tract, while a study of Australia's TIGER-like system found that 50% of the geocoded points were mapped to the wrong property parcel. The accuracy of geocoded data can also have a bearing on the quality of research that can be done using this data. One study by a group of Iowa researchers found that the common method of geocoding using TIGER datasets as described above, can cause a loss of as much as 40% of the power of a statistical analysis. An alternative is to use orthophoto or image coded data such as the Address Point data from Ordnance Survey in the UK, but such datasets are generally expensive. Because of this, it is quite important to avoid using interpolated results except for non-critical applications, such as pizza delivery. Interpolated geocoding is usually not appropriate for making authoritative decisions, for example if life safety will be affected by that decision. Emergency services, for example, do not make an authoritative decision based on their interpolations; an ambulance or fire truck will always be dispatched regardless of what the map says.

Other Techniques

In rural areas or other places lacking high quality street network data and addressing, GPS is useful for mapping a location. For traffic accidents, geocoding to a street intersection or midpoint along a street centerline is a suitable technique. Most highways in developed countries have mile markers to aid in emergency response, maintenance, and navigation. It is also possible to use a combination of these geocoding techniques — using a particular technique for certain cases and situations and other techniques for other cases. In contrast to geocoding of structured postal address records, toponym resolution maps place names in unstructured document collections to their corresponding spatial footprints.

Research

Research has introduced a new approach to the control and knowledge aspects of geocoding, by using an agent-based paradigm. In addition to the new paradigm for geocoding, additional correction techniques and control algorithms have been developed. The approach represents the geographic elements commonly found in addresses as individual agents. This provides a commonality and duality to control and geographic representation. In addition to scientific publication, the new approach and subsequent prototype gained national media coverage in Australia. The research was conducted at Curtin University in Perth, Western Australia.

Uses

Geocoded locations are useful in many GIS analysis, cartography, decision making workflow, transaction mash-up, or injected into larger business processes. On the web, geocoding is used in services like routing and local search. Geocoding, along with GPS provides location data for geotagging media, such as photographs or RSS items.

Privacy Concerns

The proliferation and ease of access to geocoding (and reverse-geocoding) services raises privacy concerns. For example, in mapping crime incidents, law enforcement agencies aim to balance the privacy rights of victims and offenders, with the public's right to know. Law enforcement agencies have experimented with alternative geocoding techniques that allow them to mask a portion of the locational detail (e.g., address specifics that would lead to identifying a victim or offender). As well, in providing online crime mapping to the public, they also place disclaimers regarding the locational accuracy of points on the map, acknowledging these location masking techniques, and impose terms of use for the information.

Raster Data Spatial Analysis

Local Operations

Local functions process a grid on a cell-by-cell basis, that is, the output value of each cell depends on the values of corresponding cells in the rasters input for the analysis. The following are the examples of the local operations.

Arithmetic Operation

Grids can undergo a range of arithmetic operations such as addition, subtraction, multiplication, and division. If the data in grids (operands) is in the form of integer then the data in the resultant grid after any mathematical operation would also be integer.

Only in one case, when any integer is divided by zero the corresponding resultant cell will be undefined and are assigned to no data. No data cells always remain No data in arithmetic operations.

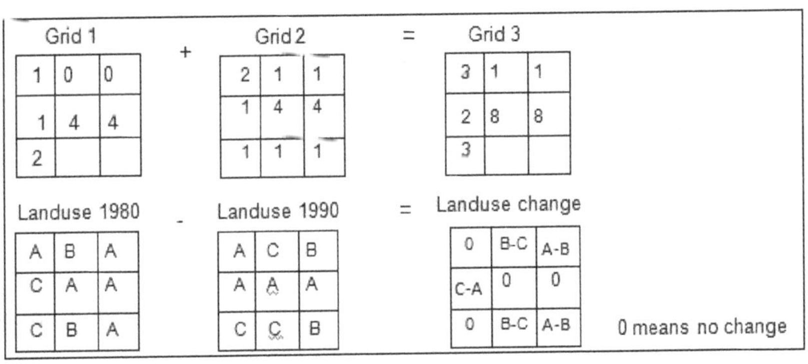

SELECT returns original cell value if the cell meets the logical expression criteria otherwise the output cell is assigned no data.

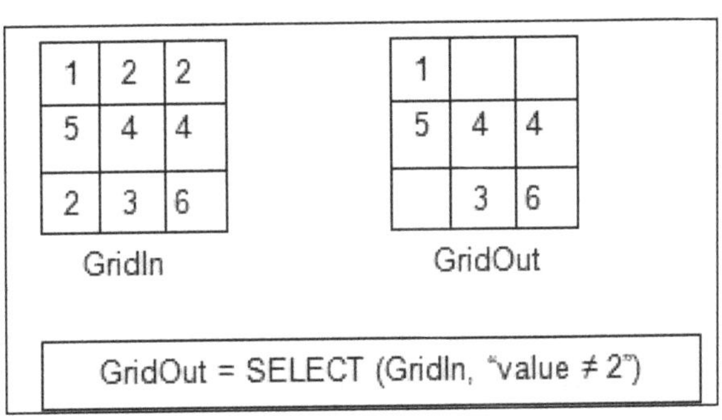

TEST performs a Boolean evaluation of the input raster using a logical expression. If the cell meets the logical expression criteria it assigns the output cell a value of 1 otherwise the output cell is assigned 0.

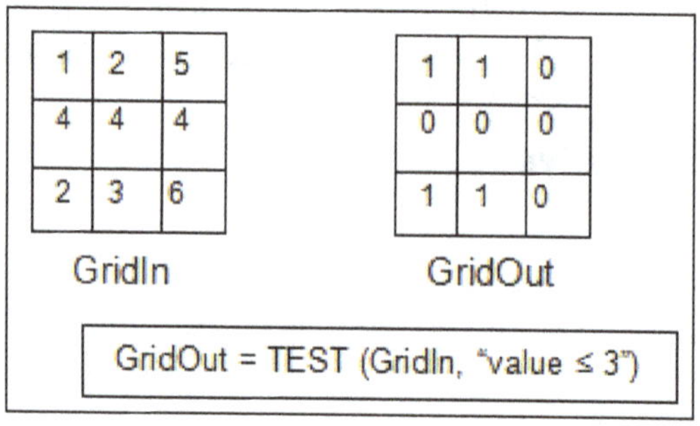

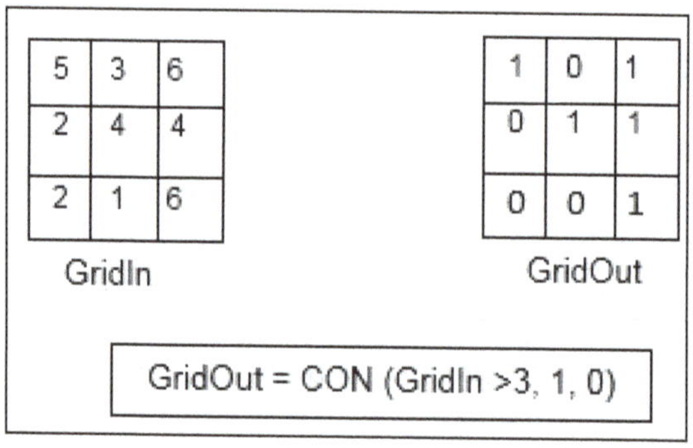

CON performs a conditional evaluation on each cell of the input raster. It tests for a user specified logical expression and return user specified values.

Focal Operations (Neighborhood Analysis)

The value of a cell in the output raster depends upon the value of the corresponding cells and their neighboring cells in the input rasters. The neighborhood for a cell is generally taken as a 3×3 matrix (window) in which the cell itself occupies the centre and is surrounded by the others eight cells. With each cell in the input getting processed, the neighborhood window keeps moving.

Spatial Aggregation

It is the process of reducing the number of cells in the raster layer to achieve generalization. It is not a compression technique because the same geographic space is represented using small number of cells of coarse spatial resolution. The choice of neighborhood window size results in different outputs. A large window size will result in higher aggregation level which implies a greater loss of details. A user can use one of the following aggregating methods:

- Averaging method: It compute average value of the cells over the window and is used as the value of the aggregated cell.

- Central cell method: Value of the cell at the centre of the window is the value of the aggregated cell.
- Median method: It computes the median value of all the cells over the window and uses it as the value of the aggregated cell.

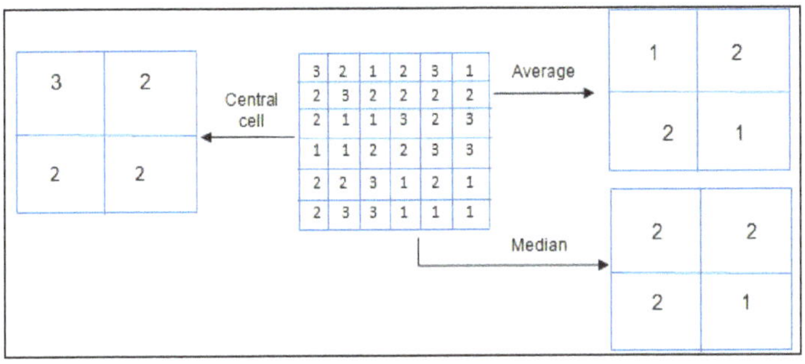

Filtering

Filtering is used to enhance an image. The value in each cell of the raster image indicates the degree of brightness at that point. The change in brightness value per unit distance for any part of the image gives rise to an image characteristic called spatial frequency. If the change is small, the area is of low frequency and if the brightness values change rapidly over small distance the area is of high frequency. The method of filtering uses a filter window or kernel (generally a 3× 3 matrix) which is passed over the whole image. On passing the kernel over the image, the coefficients in the kernels are multiplied by the value of the corresponding cells and the average value is assigned to the cell at the centre of the kernel.

Filtering suppresses noise in the image and highlight specific characteristics.

Original Image.

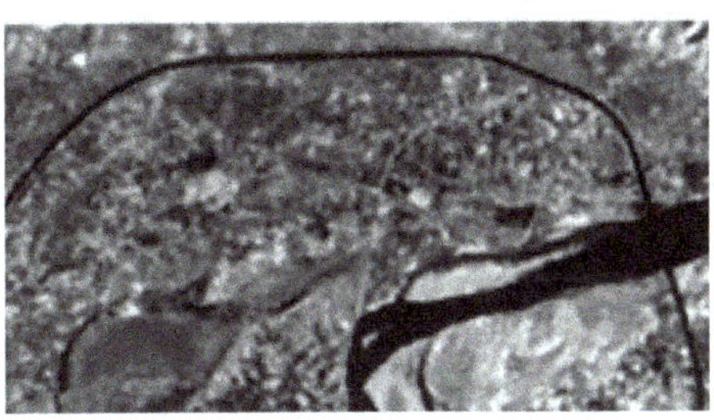

Smooth image obtained after Low pass filtering.

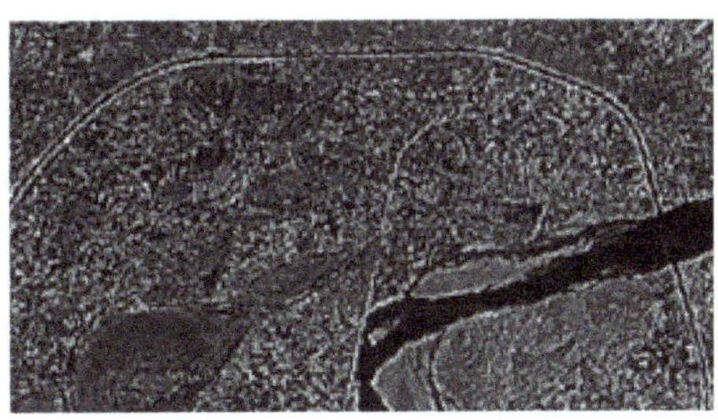

Sharpened image obtained after high pass filtering.

This neighborhood operation is used for DEMs. The angle of slope can be calculated as:

$$\tan a = \frac{vd}{hd}$$

$$a = \tan^{-1}\frac{vd}{hd}$$

Where vd is the difference in height between two points and hd is the horizontal distance between the two points for which slope is to be determined. The illustration below described the method of slope calculation.

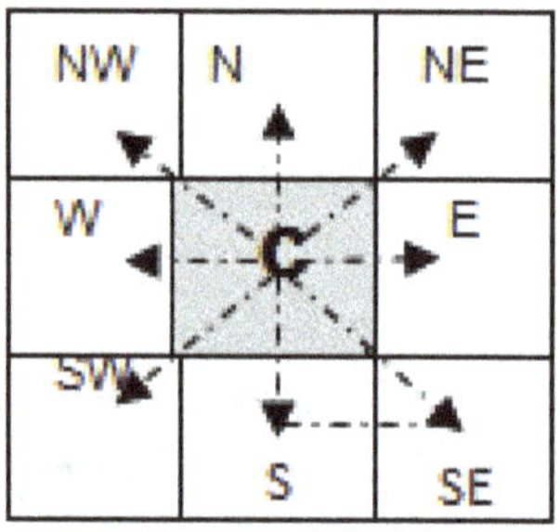

In a 3×3 window of the raster layer, the centre cell is surrounded by 8 cells and each can be given a direction. We can calculate the slope angle of the centre cell for each of these directions. The horizontal distance between the centre cell and the cells in N, E, W and S is the cell size of the raster data and the vertical distance is the difference of height between the centre cell and each of the four cells. In case of cells in NW, NE, SW and SE, the horizontal distance between the centre cell and these cells can be calculated as per Pythagoras formula:

$$hd = \sqrt{(S-C)^2 + (S-SE)^2}$$

The values of vd and hd are substituted in the equation to determine the angle of slope.

Zonal Operations

Zonal functions process grid in such a manner that cells of same zone are analyzed together. A zone may or may not be contiguous. The output value for each location depends upon the value of cell at that location and the association the location has within a zone.

Zonal geometry: The zonal geometric functions return geometric information about each zone in a grid. Following geometries can be calculated using it.

a. Area: For each zone in the input raster, Zonal area calculates the area and assign it to each cell of the zone on the output raster. The area is calculated by the number of cells that comprise the zone multiplied by the current cell size.

1	2	2	3
4	1	1	3
4	5	5	2
1	1	5	2

5	4	4	2
2	5	5	2
2	3	3	4
5	5	3	4

 # Cell size is 1

b. Perimeter: Zonalperimeter determines the perimeter of each zone on the input raster and assigns it to each cell in the zone on an output raster.

1	2	2	3
4	1	1	3
4	5	5	2
1	1	5	2

16	12	12	6
6	16	16	6
6	8	8	12
16	16	8	12

 # Cell size is 1

c. Centroid: Zonal centroid approximates the geometry of each zone by creating an ellipse fixed at the centroid of each zonal spatial shape. The area of each ellipse is equal to the area of the zone it represents.

1	1	2	2	5	5
1	2	2	2	5	5
	2	2	2	4	4
3	3	4	4	4	3
3	6	6	6	4	3
3	3	6	6	6	3

1					
		2			5
				4	
		3			
			6		

Zonal statistics: It calculates statistics for each zone of the input dataset. The zonal statistical functions return a statistical measure of the values of each zone. The measure can be the mean, median, majority, standard deviation, sum, minimum, maximum, or range of the input values.

Global Operations

The value of each cell in the output raster is a function of the entire grid. Following are few of the global functions:

REGIONGROUP: It aggregates the cells with same values into groups. It starts at upper left corner cell and proceed left to right assigning group numbers based on cells that touch and have the same cell values.

4	3	4	3
3	3	4	3
4	4	4	3
4	4	3	3

GridIn

1	2	3	4
2	2	3	4
3	3	3	4
3	3	4	4

GridOut

Value attribute table		
Value	Count	Link
1	1	4
2	3	3
3	7	4
4	5	3

The value attribute table contains link along with the values and their count. Link is the original value of the cells before the grouping occurred.

EUCDISTANCE: The Euclidean distance grid identifies the distance from each cell to the closest source cell.

1	1		
	1		
		2	

0	0	1	2
1	0	1	2
1.4	1	1	1.4
2	1	0	1

The output values for the Euclidean grid are floating-point distance values. If a cell is at an equal distance from two or more sources, it is assigned to the source that is first encountered in the scanning process (The process of scanning starts at the upper left corner and moves from left-to-right, top-to-bottom).

Other global functions include Costdistance, Costpath, Eucdirection, Slice etc.

References

- Banerjee, Sudipto; Carlin, Bradley P.; Gelfand, Alan E. (2014), Hierarchical Modeling and Analysis for Spatial Data, Second Edition, Monographs on Statistics and Applied Probability (2nd ed.), Chapman and Hall/CRC, ISBN 9781439819173
- Ratcliffe, Jerry H. (2001). “On the accuracy of TIGER-type geocoded address data in relation to cadastral and census areal units” (PDF). International Journal of Geographical Information Science. 15 (5).
- (eds.), Wolfgang Kresse, David M. Danko (2010). Springer handbook of geographic information (1. ed.). Berlin: Springer. pp. 82–83. ISBN 9783540726807. CS1 maint: Extra text: authors list (link)
- Jennifer Foreshew (24 November 2009). “Difficult addresses no problem for IntelliGeoLocator”. The Australian. Retrieved 17, March 2020
- Datta, Abhirup; Banerjee, Sudipto; Finley, Andrew O.; Gelfand, Alan E. (2016). “Hierarchical Nearest Neighbor Gaussian Process Models for Large Geostatistical Datasets”. Journal of the American Statistical Association. doi:10.1080/01621459.2015.1044091
- “Spatially enabling the data: What is geocoding?”. National Criminal Justice Reference Service. Retrieved 11, February 2020

- Banerjee, Sudipto; Gelfand, Alan E.; Finley, Andrew O.; Sang, Huiyan (2008). "Gaussian predictive process models for large spatial datasets". Journal of the Royal Statistical Society Series B. 70 (4): 825–848. doi:10.1111/j.1467-9868.2008.00663.x
- Olivares, Miriam. "Geographic Information Systems at Yale: Geocoding Resources". guides.library.yale.edu. Retrieved 21, August 2020
- Department of Education, Western Australia (April 2011). "X marks the spot". School Matters. Retrieved 13, June 2020

6

Diverse Area of GIS Applications

Geographic information system is an interdisciplinary subject which is essential for the understanding of related fields. This chapter discusses the significance of GIS in other sectors like public health, aquatic science, hydrology and environmental governance. The aim of this chapter is to explore the allied fields of geographic information system.

GIS and Public Health

Geographic information systems (GISs) and geographic information science (GIScience) combine computer-mapping capabilities with additional database management and data analysis tools. Commercial GIS systems are very powerful and have touched many applications and industries, including environmental science, urban planning, agricultural applications, and others.

Public health is another focus area that has made increasing use of GIS techniques. A strict definition of public health is difficult to pin down, as it is used in different ways by different groups. In general, public health differs from personal health in that it is (1) focused on the health of populations rather than of individuals, (2) focused more on prevention than on treatment, and (3) operates in a mainly governmental (rather than private) context. These efforts fall naturally within the domain of problems requiring use of spatial analysis as part of the solution, and GIS and other spatial analysis tools are therefore recognized as providing potentially transformational capabilities for public health efforts.

This article presents some history of use of geographic information and geographic information systems in public health application areas, provides some examples showing the utilization of GIS techniques in solving specific public health problems, and finally addresses several potential issues arising from increased use of these GIS techniques in the public health arena.

History

Public health efforts have been based on analysis and use of spatial data for many years. Dr. John Snow (physician), often credited as the father of epidemiology, is arguably the most famous of those examples. Dr. Snow used a hand-drawn map to analyze the geographic locations of deaths related to cholera in London in the mid-1850s. His map, which superimposed the locations of cholera deaths with those of public water supplies, pinpointed the Broad Street pump as the most likely source of the cholera outbreak. Removal of the pump handle led to a rapid decline in the incidence of cholera, helping the medical community to eventually conclude that cholera was a water-borne disease.

Dr. Snow's map showing cholera cases in London during the epidemic of 1854.

Dr. Snow's work provides an indication of how a GIS could benefit public health investigations and other research. He continued to analyze his data, eventually showing that the incidence rate of cholera was also related to local elevation as well as soil type and alkalinity. Low-lying areas, particularly those with poorly draining soil, were found to have higher incidence rates for cholera, which Dr. Snow attributed to the pools of water that tended to collect there, again showing evidence that cholera was in fact a water-borne disease (rather than one borne by 'miasma' as was commonly believed at the time.

This is an early example of what has come to be known as disease diffusion mapping, an area of study based on the idea that a disease starts from some source or central point and then spreads throughout the local area according to patterns and conditions there. This is another area of research where the capabilities of a GIS have been shown to be of help to practitioners.

GIS for Public Health

Today's public health problems are much larger in scope than those Dr. Snow faced, and researchers today depend on modern GIS and other computer mapping applications to assist in their analyses. For example, see the map to the right depicting death rates from heart disease among white males above age 35 in the US between 2000 and 2004.

Public health informatics (PHI) is an emerging specialty which focuses on the application of information science and technology to public health practice and research. As part of that effort, a GIS – or more generally a spatial decision support system (SDSS) – offers improved geographic visualization techniques, leading to faster, better, and more robust understanding and decision-making capabilities in the public health arena.

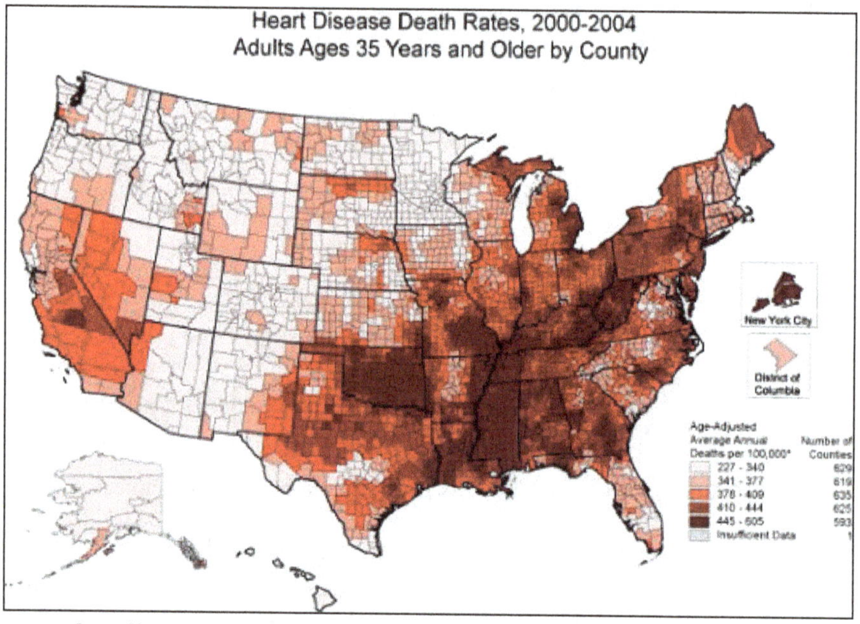

More modern disease map showing deaths from heart disease among white males in the US from 2000–2004.

For example, GIS displays have been used to show a clear relationship between clusters of emergent Hepatitis C cases and those of known intravenous drug users in Connecticut. Causality is difficult to prove conclusively – collocation does not establish causation – but confirmation of previously established causal relationships (like intravenous drug use and Hepatitis C) can strengthen acceptance of those relationships, as well as help to demonstrate the utility and reliability of GIS-related solution techniques. Conversely, showing the coincidence of potential causal factors with the ultimate effect can help suggest a potential causal relationship, thereby driving further investigation and analysis (source needed?).

Alternately, GIS techniques have been used to show a lack of correlation between causes and effects or between different effects. For example, the distributions of both birth defects and infant mortality in Iowa were studied, and the researchers found no relationship in those data. This led to the conclusion that birth defects and infant mortality are likely unrelated, and are likely due to different causes and risk factors.

GIS can support public health in different ways as well. First and foremost, GIS displays can help inform proper understanding and drive better decisions. For example, elimination of health disparities is one of two primary goals of Healthy People 2010, one of the preeminent public health programs in existence today in the US. GIS can play a significant role in that effort, helping public health practitioners identify areas of disparities or inequities, and ideally helping them identify and develop solutions to address those shortcomings. GIS can also help researchers integrate disparate data from a wide variety of sources, and can even be used to enforce quality control measures on those data. Much public health data is still manually generated, and is therefore subject to human-generated mistakes and miscoding. For example, geographic analysis of health care data from North Carolina showed that just over 40% of the records contained errors of some sort in the geographic information (city, county, or zip code), errors that would have gone undetected without the visual displays provided by GIS. Correction of these errors led not only to more correct GIS displays, but also improved ALL analyses using those data.

Issues with GIS for Public Health

There are also concerns or issues with use of GIS tools for public health efforts. Chief among those is a concern for privacy and confidentiality of individuals. Public health is concerned about the health of the population as a whole, but must use data on the health of individuals to make many of those assessments, and protecting the privacy and confidentiality of those individuals is of paramount importance. Use of GIS displays and related databases raises the potential of compromising those privacy standards, so some precautions are necessary to avoid pinpointing individuals based on spatial data. For example, data may need to be aggregated to cover larger areas such as a zip code or county, helping to mask individual identities. Maps can also be constructed at smaller scales so that less detail is revealed. Alternately, key identifying features (such as the road and street network) can be left off the maps to mask exact location, or it may even be advisable to intentionally offset the location markers by some random amount if deemed necessary.

It is well established in the literature that statistical inference based on aggregated data can lead researchers to erroneous conclusions, suggesting relationships that in fact do not exist or obscuring relationships that do in fact exist. This issue is known as the modifiable areal unit problem. For example, New York public health officials worried that cancer clusters and causes would be misidentified after they were forced to post maps showing cancer cases by ZIP code on the internet. Their assertion was that ZIP codes were designed for a purpose unrelated to public health issues, and so use of these arbitrary boundaries might lead to inappropriate groupings and then to incorrect conclusions.

Summary

Use of GIS in public health is an application area still in its infancy. Like most new applications, there is a lot of promise, but also a lot of pitfalls that must be avoided along the way. Many researchers and practitioners are concentrating of this effort, hoping that the benefits outweigh the risks and the costs associated with this emerging application area for modern GIS techniques.

GIS and Aquatic Science

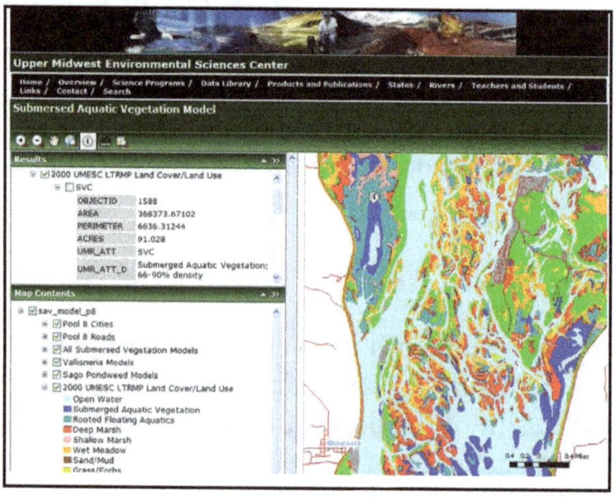

ArcGIS Server website depicting submersed aquatic vegetation.

Geographic Information Systems (GIS) has become an integral part of aquatic science and limnology. Water by its very nature is dynamic. Features associated with water are thus ever-changing. To be able to keep up with these changes, technological advancements have given scientists methods to enhance all aspects of scientific investigation, from satellite tracking of wildlife to computer mapping of habitats. Agencies like the US Geological Survey, US Fish and Wildlife Service as well as other federal and state agencies are utilizing GIS to aid in their conservation efforts.

GIS is being used in multiple fields of aquatic science from limnology, hydrology, aquatic botany, stream ecology, oceanography and marine biology. Applications include using satellite imagery to identify, monitor and mitigate habitat loss. Imagery can also show the condition of inaccessible areas. Scientists can track movements and develop a strategy to locate locations of concern. GIS can be used to track invasive species, endangered species, and population changes.

One of the advantages of the system is the availability for the information to be shared and updated at any time through the use of web-based data collection.

GIS and Fish

In the past, GIS was not a practical source of analysis due to the difficulty in obtaining spatial data on habitats or organisms in underwater environments. With the advancement of radio telemetry, hydroacoustic telemetry and side-scan sonar biologists have been able to track fish species and create databases that can be incorporated into a GIS program to create a geographical representation. Using radio and hydro-acoustic telemetry, biologists are able to locate fish and acquire relatable data for those sites, this data may include substrate samples, temperature, and conductivity. Side-scan sonar allows biologists to map out a river bottom to gain a representation of possible habitats that are used. These two sets of data can be overlaid to delineate the distribution of fish and their habitats for fish. This method has been used in the study of the pallid sturgeon.

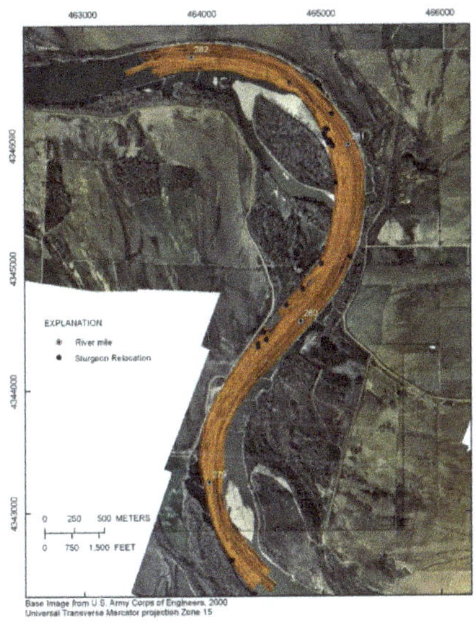

USGS sidescan radar image over base image from Army Corps of Engineers, indicating sturgeon location and river mile.

Over a period of time large amounts of data are collected and can be used to track patterns of migration, spawning locations and preferred habitat. Before, this data would be mapped and overlaid manually. Now this data can be entered into a GIS program and be layered, organized and analyzed in a way that was not possible to do in the past. Layering within a GIS program allows for the scientist to look at multiple species at once to find possible watersheds that are shared by these species, or to specifically choose one species for further examination. The US Geological Survey (USGS) in, cooperation with other agencies, were able to use GIS in helping map out habitat areas and movement patterns of pallid sturgeon. At the Columbia Environmental Research Center their effort relies on a customized ArcPad and ArcGIS, both ESRI (Environmental Systems Research Institute) applications, to record sturgeon movements to streamline data collection. A relational database was developed to manage tabular data for each individual sturgeon, including initial capture and reproductive physiology. Movement maps can be created for individual sturgeon. These maps help track the movements of each sturgeon through space and time. This allowed these researchers to prioritize and schedule field personnel efforts to track, map, and recapture sturgeon.

GIS and Macrophytes

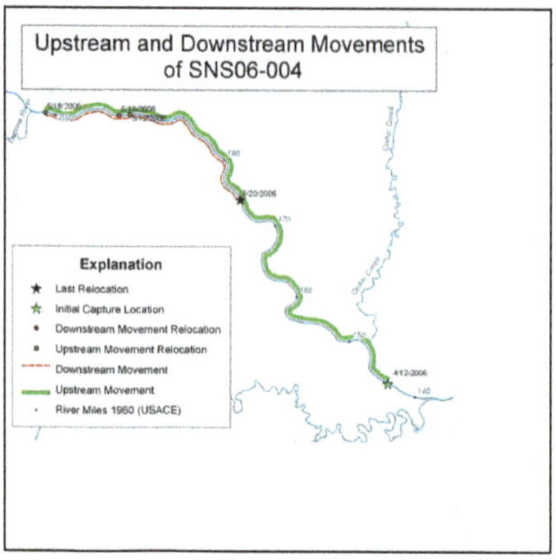

Map created from GIS database depicting the movements of individual sturgeon.

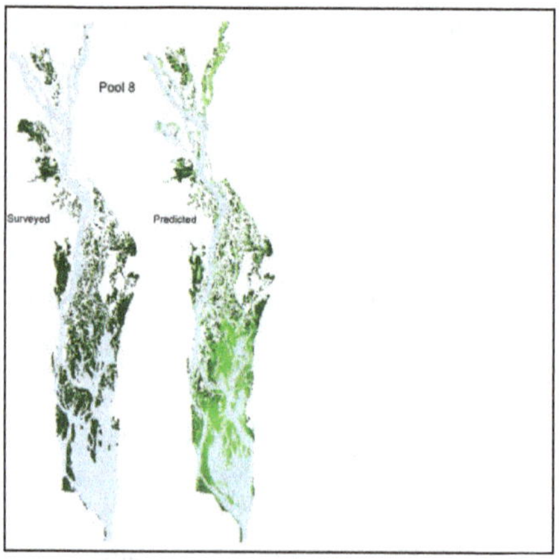

Surveyed (left) and predicted (right) distributions of submersed aquatic vegetation distribution Upper Mississippi River in 1989. The survey data were from the land cover/land use geographic information created by the U.S. Geological Survey Upper Midwest Environmental Sciences Center on the basis of interpretation of aerial photography of 1989.

Macrophytes are an important part of healthy ecosystems. They provide habitat, refuge, and food for fish, wildlife, and other organisms. Though natural occurring species are of great interest so are the invasive species that occur alongside these in our environment. GIS is being used by agencies and their respective resource managers as a tool to model these important macrophyte species. Through the use of GIS resource managers can assess the distributions of this important aspect of aquatic environments through a spatial and temporal scale. The ability to track vegetation change through time and space to make predictions about vegetation change are some of the many possibilities of GIS. Accurate maps of the aquatic plant distribution within an aquatic ecosystem are an essential part resource management.

It is possible to predict the possible occurrences of aquatic vegetation. For example, the USGS has created a model for the American wild celery (Vallisneria americana) by developing a statistical model that calculates the probability of submersed aquatic vegetation. They established a web link to an Environmental Systems Research Institute (ESRI) ArcGIS Server website *Submersed Aquatic Vegetation Model to make their model predictions available online. These predictions for distribution of submerged aquatic vegetation can potentially have an effect on foraging birds by creating avoidance zones by humans. If it is known where these areas are, birds can be left alone to feed undisturbed. When there are years where the aquatic vegetation is predicted to be limited in these important wildlife habitats, managers can be alerted.

Invasive species have become a great conservation concern for resource managers. GIS allows managers to map out plant locations and abundances. These maps can then be used to determine the threat of these invasive plants and help the managers decide on management strategies. Surveys of these species can be conducted and then downloaded into a GIS system. Coupled with this, native species can be included to determine how these communities respond with each other. By using known data of preexisting invasive species GIS models could predict future outbreaks by comparing biological factors. The Connecticut Agricultural Experiment Station Invasive Aquatic Species Program (CAES IAPP) is using GIS to evaluate risk factors. GIS allows managers to georeference plant locations and abundance. This allows for managers to display invasive communities alongside native species for study and management.

GIS and Hydrology

Geographic information systems (GISs) have become a useful and important tool in hydrology and to hydrologists in the scientific study and management of water resources. Climate change and greater demands on water resources require a more knowledgeable disposition of arguably one of our most vital resources. As every hydrologist knows, water is constantly in motion. Because water in its occurrence varies spatially and temporally throughout the hydrologic cycle, its study using GIS is especially practical. GIS systems previously were mostly static in their geospatial representation of hydrologic features. Today, GIS platforms have become increasingly dynamic, narrowing the gap between historical data and current hydrologic reality.

The elementary water cycle has inputs equal to outputs plus or minus change in storage. Hydrologists make use of a hydrologic budget when they study a watershed. A watershed is a spatial area, and the occurrence of water throughout its space varies by time. In the hydrologic budget are inputs such as precipitation, surface flows in, and groundwater flows in. Outputs are evapotranspiration, infiltration, surface runoff, and surface/groundwater flows out. All of these quantities, including storage, can be measured or estimated, and their characteristics can be graphically displayed in GIS and studied.

As a subset of hydrology, hydrogeology is concerned with the occurrence, distribution, and movement of groundwater. Moreover, hydrogeology is concerned with the manner in which groundwater is stored and its availability for use. The characteristics of groundwater can readily be input into GIS for further study and management of water resources. Because 98% of the world's available freshwater is groundwater, the need to keep a closer eye on its disposition is readily apparent.

GIS in Surface Water

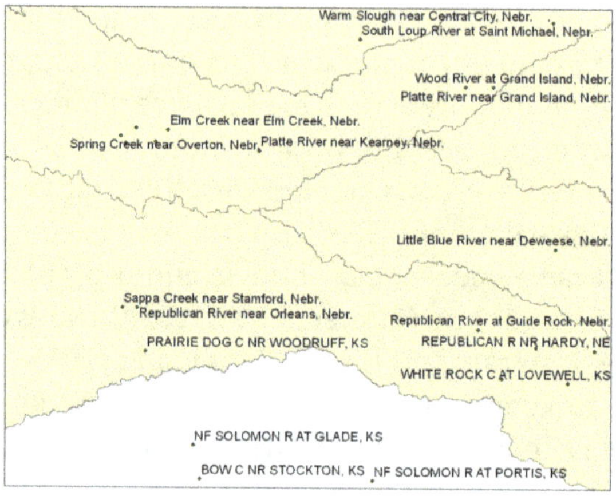

USGS Real-Time streamflow gage locations with hyperlinks within a GIS to the data.

It is possible to access historical and real time streamflow data via the Internet. Embedded within a GIS are layers with stream locations and gage or measuring/monitoring sites. It's also possible to link radio transmitted and remotely sensed (Remote Sensing) data in GIS. Historical and real time data are available from the United States Geological Survey (USGS) in the form of gage height and streamflow or discharge in cubic feet per second. Within a GIS, it's possible to direct link via the Internet to real time data. Other sources of data for flood information and water quality come from the National Weather Service (NWS) and United States Environmental Protection Agency (EPA). All these data are available for analysis within GIS, providing a spatial representation of what would otherwise be data in a table type format.

GIS is much more capable of displaying data spatially than temporally. Within one GIS, ESRI's ArcGIS for example, is it possible to delineate a watershed. Digital elevation model (DEM) data are layered with hydrographic data so that the boundaries of a watershed may be determined. Watershed delineation aids the hydrologist or water resource manager in understanding where runoff from precipitation or snowmelt will eventually drain. In the case of snowmelt, snowpack coverage may be determined from ground stations or remotely sensed observers and input into GIS to determine or predict how much water can be counted on to be available for use by cities, agriculture, and environmental habitat.

Another useful application for GIS regards precipitation, but other hydrologic data (evapotranspiration, infiltration, and groundwater) may be treated similarly. Precipitation is an area event measured using data from point locations. The difficulty in using point data lies in extrapolating these point measurements to areas. One useful method to extrapolate data is to construct Thiessen polygons which assess the distance and geometry of points in a plane and determines representative areas for which to assign precipitation values. GIS applications like ArcGIS are capable of constructing Thiessen polygons, and other methods of determining area precipitation are viable with GIS as well.

A step up in complexity from manual analysis of select spatially depicted hydrologic data is to display a representative version of hydrologic reality and perhaps merge it with a numerical or other model which might predict what might happen say x amount of rainfall occurs or to forecast, for example, runoff following the passage of an approaching weather system. One such method to do this would be to connect a GIS data model with a simulation model. The GIS data model has all the relevant surface water features with attributes that describe historical or current hydrologic data. The data model structures all the pertinent data to arrive at a representative depiction of hydrologic reality for display and analysis. One data model which does this is Arc Hydro, created cooperatively by ESRI and the Center for Research in Water Resources (CRWR) at the University of Texas at Austin to work within ESRI's ArcGIS. It is important to understand the data model does not predict as this is the function of the simulation model that Arc Hydro might feed. The simulation model is very complex and beyond the scope of this article.

By synthesizing GIS technology with hydrologic data, it has become possible to elucidate the effects of watershed-scale land-use and land-cover changes. For example, with growing pressures on water resources there is a strong interest in how forestation affects water yields. GIS and remote sensing facilitate quantifying long-term changes in forest cover since aerial photography records are available across much of the United States since as early as the 1930s. Even earlier than the 1930s the USGS started systematically gauging many watersheds throughout the country. Once long-term land-cover trends have been quantified in a gauged watershed, it becomes possible to statistically compare the long-term land-cover changes with the land-use changes to determine, for example, if forestation is actually reducing streamflow as is widely perceived. Thus using GIS data together with hydrology data can allow for knowledge based water resources decision making at far lower costs than traditional methods!

GIS in Groundwater

As mentioned earlier, 98% of the available freshwater (negating polar and glacial ice) for human and environmental uses is in groundwater. In the United States, about 1/4 of the water used for personal, commercial/industrial, and irrigation uses comes from groundwater. With increasing demands placed on surface water resources, it is likely the demand for groundwater will increase. In some places, this resource has already been severely tapped, and even mismanaged. An example here is the surface water decline in the Republican River watershed of Nebraska and Kansas where over-pumping of groundwater for irrigation in Nebraska has depleted surface water available for downstream flow and use in Kansas resulting in a lawsuit by that state against the state of Nebraska. Although not as apparent as surface water flow, groundwater can also be characterized spatially in a GIS and analyzed by scientists and natural resource managers.

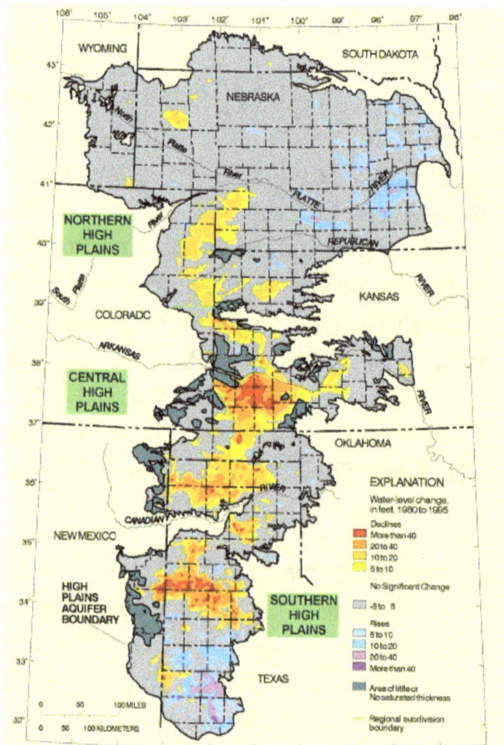

Groundwater level change of the High Plains Aquifer, 1980–95.

It can be argued that the depiction of groundwater is an even more complex task than that of surface water. The two resources are by no means disjoint, as knowing where surface water recharges groundwater and where groundwater flows supply surface water is an important aspect of the hydrologic cycle. Hydrogeology is especially well suited to GIS. Groundwater moves much more slowly than surface water, on the order of less than a meter per day up to perhaps a hundred meters per day, and is 3-dimensional in flow. In contrast, surface water flows much faster and is more two-dimensional. Groundwater flow is a function of geology and "head," the total potential energy at a location. Groundwater flows from higher head to lower head at a travel rate and flow path dictated by geology. Head values, geology, groundwater flow direction, even water table height and location of aquifers are among the quantities which may be presented spatially in GIS and used for analysis, management of water availability and water quality, and land use practices.

A very large amount of data from wells is available such as location, depth to water, stratigraphy, water quality and chemistry, aquifer characteristics, and the list goes on. The volume of data can be managed in a GIS and manipulated to display spatial characteristics for analysis and water resource planning. For example, in a simple application of GIS, the effect of a new well can be studied on the existing groundwater and surface water. The results of such a study can be used by decision makers to determine whether or not to proceed with drilling.

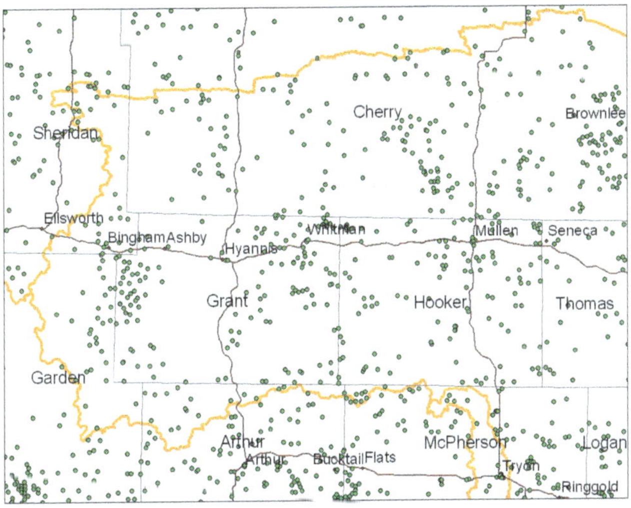

Nebraska Sandhills: registered well locations in the Upper Loup basin.

An especially useful application of GIS concerns water quality in groundwater. For construction/situating of industrial plants, landfills, agricultural activities, and other potential groundwater contamination sources, it is useful to know how existing groundwater supplies could be affected or would be at risk of impact. Further, in the case of groundwater contamination and the need for subsequent containment and cleanup of the contaminant, an existing framework of the groundwater system would be valuable in planning remediation measures. This GIS could be the front end to a groundwater modeling simulation devised to fully capture the contaminant. An additional example concerning the use of GIS addresses a common problem associated with groundwater pumping and land subsidence or intrusion in coastal areas. Areas that have been overpumped of groundwater can subside, and when near the sea, this may invite flooding. Also, overpumping of groundwater in coastal regions may bring a different problem, such as the case in California where salt-water intrusion has compromised the aquifer. Generally, a salt water interface inland of the coast extends below the land surface dependent on the distance from the coast. Overpumping can bring the salt water interface to a higher position and contaminate an aquifer. A careful study and management of groundwater within GIS or with modeled GIS data can forestall or alleviate these problems.

GIS and Environmental Governance

U.S. Federal Emergency Management Agency (FEMA) Administrator W. Craig Fugate speaking at a Red Cross seminar on using social media during natural disasters. GIS has an integral role to play in such agendas.

Geographic Information Systems (GIS) is a commonly used tool for environmental management, modelling and planning. As simply defined by Michael Goodchild, GIS is as 'a computer system for handling geographic information in a digital form'. In recent years it has played an integral role in participatory, collaborative and open data philosophies. Social and technological evolutions have elevated 'digital' and 'environmental' agendas to the forefront of public policy, the global media and the private sector.

Government departments routinely use digital spatial platforms to plan and model proposed changes to road networks, building design, greenbelt land, utility provision, crime prevention, energy production, waste management and security. Non-profit organizations also incorporate geospatial and web-mapping approaches into political campaigns to lobby governments, to protest against socially or environmentally harmful companies, and to generate public support. Private business, whether in land management, resource extraction, retail, manufacturing or social media for example, also incorporate GIS into overall profit-making strategies.

Citizen Science and GIS

Citizen science is part of the wider emphasis upon public involvement in expert fields across Western democracies. The term is 'often used to describe communities or networks of citizens who act as observers in some domain of science'. Although more narrowly used to describe the shift to specifically user-generated forms of knowledge creation, it has been routinely invoked in both the public participatory GIS and environmental governance literature at large.

The Secretary of the US Navy, Ray Mabus is briefed on the Deepwater Horizon oil spill response. A web-based GIS is visible in the background. The NOAA-developed Environmental Response Management Application (ERMA) was designed to assist resource managers post-spill.

National Audubon Society: Gulf Spill Bird Tracker

Much of the citizen science literature is grounded in wildlife study. For example, Goodchild references the National Audubon Society's Christmas Bird Count (CBC) as a classic case of citizen science in action. Each year over the winter period, the American conservation organization encourages volunteer bird-watchers to gather information on the number of bird species in their local area. Once field data has been collected, each volunteer is able to submit their bird sightings into an online database, for the benefit of both scientific researchers and bird enthusiasts. The eBird project – enabling the general public to explore a range of map- and chart-based bird species datasets - is a result of these yearly mass volunteer events.

Of particular interest is the Gulf Spill Bird Tracker; an interactive sightings map for ten species deemed at risk from the Deepwater Horizon oil spill in 2010. Gulf Coast bird watchers were encouraged (at the time when it was live) to submit their sightings of a range of at-danger birds (such as the brown pelican, roseate spoonbill and the Wilson's plover), to help aid the clean-up operation, and pin-point beaches most affected by the oil spill. The National Audubon Society has been deeply involved in the Gulf Oil Response since the disaster, and has a dedicated program to co-ordinate resources, liaise with local government, and deploy equipment post-spill. Their 6-month report brought together some of these key factors. Not only was the National Audubon Society's citizen science initiative highlighted as the 'backbone...for understanding the [immediate] impact of the disaster', but also for long-term efforts to monitor the health of imperilled species in the Gulf Coast region. Moreover, their grassroots ethos has mobilized a vast number of Gulf volunteers to 'urge elected officials and government agencies to hold polluters like BP accountable', for the financial, environmental, economic and social costs associated with such disasters. This is perhaps the most obvious example of web-based mapping software (a more 'citizen-friendly' form of GIS) and environmental governance discourses colliding head on. The notion of volunteered, user-generated, citizen data is the guiding mantra for such projects, and the cornerstone of any wider attempts to lobby national governments, engage with local community groups, and generate scientific research.

Mapping for Change

Another example of citizen science and GIS in action is taken from inside the academy. University College London (UCL) and London 21 sustainability network's Mapping for Change initiative has encouraged voluntary groups, local authorities and development agencies to build map-based projects to support political, social and environmental aims. They even provide a noise mapping toolkit on the Mapping for Change website itself, designed to help local communities gather evidence of intrusive and unwanted environmental nuisances and hazards. The Royal Docks community in London has used such a toolkit to help present their concerns to the Greater London Authority Environment Committee over plans to expand London City Airport. Armed with sound meters, survey sheets and access to an online mapping platform, residents were able to monitor noise levels; from overhead planes and passing motor vehicles, to birdsong and ambient river sounds.

A plane lands at London City Airport. Communities in the area have used web-based noise maps to demonstrate their objections to proposed flight expansion.

Their data was then visualized in various formats to help advance their argument. Royal Docks' residents are continually plagued by planes taking-off and landing at London City Airport, and plans to expand the number of flights a year by 50% (up to 120,000) were opposed by local communities on the basis that it would decrease their quality of life.

GIS and citizen science go hand-in-hand. Web-based mapping platforms serve as useful tools for national conservation societies, local community groups and planning departments to compile tangible data on environmental issues. Voluntary, grassroots approaches can help compile lay knowledges that feed back into more formal political frameworks.

Environmental Justice

At a local level, GIS has been frequently used to engage stakeholders in the planning of environmentally 'bad' sites. Nuclear power stations, wind farms, landfill sites, and other energy facilities are often subject to NIMBY opposition for aesthetic, health and social reasons. This is despite of their capacity to produce 'good' economic factors or employment opportunities. GIS has thus found itself deployed alongside broader cost-benefit analysis (CBA), and multi-criteria decision analysis (MCDA) approaches to socio-political conflict. Environmental Justice (EJ) activists believe these decisions act to further embed racial and class divides. GIS provides an important angle to the EJ movement.

Elements of Justice

Broadly, the EJ movement is a loose connection of social groups, stakeholders and activists who have sought to contest socio-political injustices. Commonly, this has been through a single motive; the equal distribution of environmental goods and bads. As Schlosberg contends, 'the issue of distribution is always present and always key' to the guiding EJ ethos. Yet, other demands are frequently put forward. Following Schlosberg, there are two further demands that constitute the EJ movement than a mere '[re-]distribution of environmental ills and benefits. The first is the 'recognition of the diversity of participants and experiences in affected communities'. Thus, EJ demands that people affected by environmental injustices are appropriately noticed by others. A lack of recognition in local community discourses, 'demonstrated by various forms of insults, degradation, and devaluation...', marginalize those already least able to contest political decisions. The second is the notion of *participatory justice*. According to Schlosberg;'[i]f you are not recognised, you do not participate.' Thus, recognitional justice leads directly to participatory justice. In EJ terms, participation is about involving those outside the typical political/institutional order. Democratic and participatory decision-making procedures are both an element of, and a condition for, social justice. Simultaneously, institutionalised exclusion, social cultures of misrecognition, and current distributional patterns can be challenged.

An aerial photograph of New Orleans after Hurricane Katrina in 2005. The High magitude flood event disproportionately affected the city's poor, black neighbourhoods.

Flood Hazards, Race, & Environmental Justice in New York

Maantay and Maroko's research is intended to help hazard management and disaster planning before and after a high magnitude flood event. By using a Cadastral-based Expert Dasymetric System (CEDS) they were able to estimate the number of 'vulnerable sub-populations' in the densely populated New York City area. Their research broadly supports an EJ approach to natural disaster mitigation. By highlighting the importance of equality issues and the disproportionate exposure some people have to such events, they invoke the EJ movement's notion of distributional justice. The lack of 'strong social, financial, or political support structures' are constituent factors in how people deal with large-scale disasters. The criticism of the US government's response to the devastating effects of Hurricane Katrina in 2005, draw on such notions. The online publication set-up by the Social Science Research Council (SSRC) entitled Understanding

Katrina helps to ground this research in socio-political approaches to so-called 'natural' disasters.

Maantay and Maroko believe that GIS can have an important role in these 'risk-framed' understandings. As New York is a 'hyper-heterogeneous urban area', traditional administrative population data is insufficient for flood zone/population risk research. A more fine-tuned analysis is possible if tax-level datasets (based on smaller, residential units) are used instead. Maantay and Maroko use Federal Emergency Management Agency (FEMA) floodplain and tax-level datasets to determine the potential number of people at risk. A CEDS approach significantly increases the number of affected people within New York City. In using a different GIS method, Maantay and Maroko are able to better represent the impact of such flood events on minority populations. As such, their research supports all of Schlosberg's notions of distributional, recognitional and participatory justice. As noted by themselves; 'the disadvantages suffered by racial and ethnic minority communities during and after disasters are due primarily to their low economic status and lack of political power'. Their research broadly supports the aims of the EJ movement.

New Urban Landscapes

GIS has also had a role in formulating new urban landscapes. Planned cities - designed entirely from scratch - routinely use digital technologies to visualize and demonstrate urban layouts, building structures and transport arrangements. Although CAD/CAM technologies are often used to assist in the visualization, construction, and delivery of certain engineering features, GIS helps to realize distinctly spatial components of the city. Environmental narratives of a 'carbon-free' and sustainable future favour those in the GIS industry. 'The challenge of the 21st century' as ESRI would have it, 'is to arrest the progress of climate change'. Geospatial software has played its part in developing this narrative.

Masdar City

Masdar City is a 'sustainable, zero-carbon, zero-waste' project currently under construction in the United Arab Emirates (UAE). Situated in the Abu Dhabi emirate, Masdar is described as 'an emerging global hub for renewable energy and clean technologies'. The Abu Dhabi Future Energy Company have funded and overseen the $18bn dollar project. No cars are allowed on its streets, energy is produced in part by renewable sources, building materials are 'sustainable' and water-use is controlled. GIS is being employed to plan, facilitate and test a plethora of environmental phenomena and technological processes.

A dedicated GIS team is responsible for 'managing the overall spatial information needs' of the project, starting with the drawing of a common base map with which to support the city's infrastructure. Without a spatial plan of Masdar's operative mechanisms, the city will fail to deliver its grand ambition. In particular, GIS is being used to visualize, analyse and model *land-use* in the city. Masdar – unlike any other city – has to incorporate a wealth of energy-related facilities within its perimeter. As EJ activists are all too aware, the siting of such facilities can be a key area of conflict. Masdar's water treatment and sewage plants, material recycling centre, solar power plant, geothermal test site, solar panel test field and concrete batching plant all need to be situated inside the city's boundaries. As CH2M HILL's Site Control and GIS Manager for the Masdar project confirms; 'never have so many environmental facilities come together in one place'. GIS is the central tool with which to imagine – in a digital environment – different siting scenarios. In this case, GIS is seen to operate as a decision-making tool; informing the practitioners who work on the Masdar project.

GIS is also being used to model some of Masdar's key infrastructural features directly. Its involvement in simulating the city-wide Personal Rapid Transport System (PRTS) is one such example. As common road vehicles are banned from the city, the 'driverless' transit system will transport people and freight across the 7km2 area. GIS is capable of modelling the system route, due to comprise 85 passenger stations and approximately 1,700 automated vehicles. By drawing spatial buffer zones around potential PRTS stops, passenger-distance maps can visualize residential areas that fall outside of ideal service requirements. GIS is an instrumental tool in visualizing such problems. A smooth, functioning PRTS is a central infrastructural aspect of Masdar's grand vision, and engineering companies who specialize in GIS technologies have helped in realizing this digitally-orchestrated dream. Yet journalists in particular have been sceptical. As Bryan Walsh has argued; [wi]ll Masdar City ever really develop the authenticity of a real city?'. Or as Jonathan Glancey contends, will its 'ultra-modern aspects...prove to be a mirage'?

Post-political Agendas

'Post-politics' is a neologism for the consensual, participatory and techno-managerial approach to modern governance. Originally coined by Slavoj Zizek, and discussed by the likes of Jacques Rancière and Erik Swyngedouw, the post-political critique argues that life in the Western world is routinely characterized by the de-politicizing effects of a 'consensual police order'. A number of different techno-managerial 'fixes' have been sought by neoliberal governments in order to solve expressly environmental problems, rather than due political processes. As Swyngedouw has argued, such forces have 'replaced debate, disagreement and dissensus with a series of technologies of governing that fuse around consensus, agreement, accountancy merits and technocratic environmental management'. Thus if the rise of the post-political order is due to the increasing reliance upon 'technocratic environmental management', as Swyngedouw has argued, then GIS – as a tool for neoliberal environmental governance – is implicit in such an order.

The 2009 Copenhagen Summit failed to produce a legally binding treaty on national CO2 levels, despite (or perhaps because of) a near-global delegation. An example of Jacques Ranciere's partition of the sensible in action. IPCC data, based on GIS and environmental modelling processes, provide the basis for these proposed legal frameworks.

The Police, the Political and Politics

Firstly, it serves to elucidate upon claims to a 'post-political impasse'. This is best understood through what Ranciere calls *the police*, *the political* and *politics*. It is within these three terms that Ranciere carves out what he calls the true meaning of political action, and of what it is to exercise political right. As Panagia neatly summarizes; 'politics is the practice of asserting one's position that ruptures the logic of arche'. Politics is about rediscovering the art of debate, conflict and struggle, and not merely about re-organizing the administrative framework of existing political structures (namely, the state apparatus). *The political* – in Ranciere's words – is for '[t]he one who is 'unaccounted-for', the one who has no speech to be heard'. Democracy does not work towards an 'idealized-normative condition' of equal rights, but is built upon the very ontological notion of such. *Politics* brings the political into the foreground; rendering that which was previously noise, legitimate speech. Much of what is thought to be politics in the contemporary world is actually subsumed within *the police*. Policy is not politics is this sense, then. This is what Ranciere calls *the partition of the sensible*, or the established order of things.

The post-politics of contemporary governance brings all that it can into this order. Those who were previously cast outside the police structure are now "responsible' partners' in a stakeholder-based arrangement. All views that were previously antagonistic and conflictual are now brought together in a more homely, consensual arena. No room is made for 'irrational' demands. As such, Ranciere says that, '[c]onsensus is the reduction of politics to the police'. Nowhere has this been more visible than within the apocalyptic climate narratives told *ad infinitum* by environmental practitioners, policy-makers and non-governmental organizations. Consensual politics has found its home in the environmental arena.

Post-political GIS?

In short, GIS works as a tool for mediating and diffusing socio-environmental conflicts. It does so by working within Ranciere's notion of the partition of the sensible. Whilst it may allow previously unheard voices to gain a voice (in environmental justice campaigns, foremostly), it still does not – as a tool of neoliberal governance – make room for those who are deemed 'outside', unruly or conflictual to have a voice. For example, Elwood laments the notionally 'participatory' flag-waving carried out by those involved in urban GIS-based projects. As she says; 'the skills and financial and temporal costs of using GIS effectively bar many individuals, social groups and organizations from participation in research and decision-making where it is used', denying those without the means to participate, from participation. GIS does not necessaily facilitate involvement for all.

Moreover, GIS is limited in its ontological scope, reducing all things spatial to a *calculable order*. As Leszczynski contends, GIS operates a 'discourse populated by discrete objects of knowledge'; differentiating 'between the binary of truth and error'. GIS is thus a central *polic-ing* tool for contemporary socio-environmental governance. It works to order space into discrete and ordered formats. As Jeremy Crampton points out, 'the basic model of the world in GIScience texts is: points, lines, areas, surfaces and volumes', ill prepared to deal with non-discrete, continuous environmental phenomena. All that does not fit into this order is left aside. GIS thus excludes that which cannot be 'structured accordingly', excluding epistemologies that do not neatly fit into the formal computing framework; a formal framework synonymous with 'the accountancy calculus of risk and the technologies of expert administration'. The apparatus of choice for the techno-enviro-managerialist, if Ranciere and Swyngedouw's analyses are to be put forward, and Elwood, Crampton and Leszczynski's criticism's are to be accepted, is GIS.

References

- Panagia, Davide (2001). "Ceci n'est pas un argument: An Introduction to the Ten Theses". Theory and Event. 5 (3). Retrieved 04, February 2020
- McElvaney, Shannon. "Masdar City Development Programme: Using GIS Technologies to Help Plan and Build a Sustainable City" (PDF). Retrieved 11, June 2020
- Glancey, Jonathan (10 May 2011). "Inside Masdar City: a modern mirage". London: The Guardian. Retrieved 24, April 2020
- Goodchild, Michael F. (July 2009). "GIScience and Systems". International Encyclopedia of Human Geography: 526–538. doi:10.1016/B978-008044910-4.00029-8. Retrieved04, March 2020

- Crampton, J. W. (2011). "Cartographic calculations of territory". Progress in Human Geography. 35 (1): 92–103. doi:10.1177/0309132509358474
- Audubon. "Audubon Science: Christmas Bird Count" (PDF). National Audubon Society. Retrieved 23, August 2020

Permissions

All chapters in this book are published with permission under the Creative Commons Attribution Share Alike License or equivalent. Every chapter published in this book has been scrutinized by our experts. Their significance has been extensively debated. The topics covered herein carry significant information for a comprehensive understanding. They may even be implemented as practical applications or may be referred to as a beginning point for further studies.

We would like to thank the editorial team for lending their expertise to make the book truly unique. They have played a crucial role in the development of this book. Without their invaluable contributions this book wouldn't have been possible. They have made vital efforts to compile up to date information on the varied aspects of this subject to make this book a valuable addition to the collection of many professionals and students.

This book was conceptualized with the vision of imparting up-to-date and integrated information in this field. To ensure the same, a matchless editorial board was set up. Every individual on the board went through rigorous rounds of assessment to prove their worth. After which they invested a large part of their time researching and compiling the most relevant data for our readers.

The editorial board has been involved in producing this book since its inception. They have spent rigorous hours researching and exploring the diverse topics which have resulted in the successful publishing of this book. They have passed on their knowledge of decades through this book. To expedite this challenging task, the publisher supported the team at every step. A small team of assistant editors was also appointed to further simplify the editing procedure and attain best results for the readers.

Apart from the editorial board, the designing team has also invested a significant amount of their time in understanding the subject and creating the most relevant covers. They scrutinized every image to scout for the most suitable representation of the subject and create an appropriate cover for the book.

The publishing team has been an ardent support to the editorial, designing and production team. Their endless efforts to recruit the best for this project, has resulted in the accomplishment of this book. They are a veteran in the field of academics and their pool of knowledge is as vast as their experience in printing. Their expertise and guidance has proved useful at every step. Their uncompromising quality standards have made this book an exceptional effort. Their encouragement from time to time has been an inspiration for everyone.

The publisher and the editorial board hope that this book will prove to be a valuable piece of knowledge for students, practitioners and scholars across the globe.

Index

O

P

R

S

T

U

V

W

www.ingramcontent.com/pod-product-compliance
Lightning Source LLC
LaVergne TN
LVHW080622210926
848485LV00003B/200

* 9 7 8 1 6 3 9 8 9 4 3 7 6 *